Vegetable Crops

Novel Technologies

THE EDITORS

Dr. Dalasanuru Chandregowda Manjunathagowda, ARS, Scientist (Vegetable Science). Presently working at ICAR-Directorate of Onion and Garlic Research, Pune, Maharashtra, India. His area of interest is Vegetable Science includes Vegetable Genetics and Breeding, Vegetable Production and its Seed Production.

Dr. Raman Selvakumar, ARS, Scientist (Vegetable Science). Presently working at ICAR-Indian Agricultural Research Institute, New Delhi, India. His expertise in Vegetable Genetics and Breeding, and Post Harvest Technology.

Dr. Muddappa Anjanappa, Professor, Head, Department of Vegetable Science College of Horticulture, Kolar, University of Horticultural Sciences, Bagalkot. His area of interest is Vegetable Science includes Vegetable Production and Improvement.

Vegetable Crops
Novel Technologies

– *Editors* –

D.C. Manjunathagowda

R. Selvakumar

M. Anjanappa

2022

Daya Publishing House®
A Division of
Astral International Pvt. Ltd.
New Delhi – 110 002

ISBN: 978-93-90435-12-8 (HB)

Published by : **Daya Publishing House**®
A Division of
Astral International Pvt. Ltd.
– ISO 9001:2015 Certified Company –
4736/23, Ansari Road, Darya Ganj,
New Delhi-110 002
Ph. 011-43549197, 23278134
e-mail: info@astralint.com
Website: www.astralint.com

भाकृअनुप–प्याज एवं लहसुन अनुसंधान निदेशालय
राजगुरुनगर, पुणे– 410 505, महाराष्ट्र (भारत)
ICAR-DIRECTORATE OF ONION AND GARLIC RESEARCH
RAJGURUNAGAR, PUNE- 410 505, MAHARASHTRA (INDIA)
(ISO 9001:2008)
☎ (02135) 222026, 222697 Fax (02135) 224056
Email : director.dogr@icar.gov.in, Website : dogr.res.in

ICAR

DOGR

Dr. Major Singh
Director

The world's most urgent need is to increase the production of nutritious food in a sustainable manner and improving farm family income in order to ensure household food security, while at the same time conserving the natural resource base. Vegetables are vital source of mineral, vitamins and dietary fibers and play an important role in human nutrition in supplying certain constituents in which other food materials are deficient. A large number of studies have shown that the consumption of vegetables is inversely associated with the risk of cancer, particularly of alimentary canal and respiratory tract. The vegetable scenario of the country is changing rapidly both in terms of production and productivity covering 41.3 per cent of total horticulture cropped area and contributing 59.3 per cent of production. The production of vegetables has increased from 101.2 mt to 184.4 mt since from 2002-05 to 2017-18, thus it is an indicative facts of growing demand for vegetables, may be due to enhanced output per investment and access to the information of knowledge, further improve the socio-economic conditions of Indian citizens by vegetable crops cultivation.

The prospects of enhancing vegetable production by increasing area and as well as increasing productivity become absolutely essential, by adoption of modern technologies in Vegetable Science (Protected and Precision Farming Techniques) and development of new Varieties, Hybrids and Genetic Resources has become essential to enhance productivity. In this context, the book **Vegetable Crops: Novel Technologies** edited by Manjunathagowda, DC, Selvakumar, R. and Anjanappa, M provides valuable information. Such a commendable work comprehensively deals with basic and applied aspects on Vegetable Crop Science.

I am sure that this book will be of immense value to the Graduate and Post-Graduate Students, Researchers, Professors and Scientists involved in Teaching Research, and Extension activities on vegetable crops.

(Major Singh)

Preface

Vegetables are cheapest nutrient-rich food in human diet, as they consist of carbohydrates, proteins, fats, minerals, vitamins, fibres and phytochemicals. The value of consuming vegetables is known all over the world, besides medicinal values. Vegetables play an important role in food protection and rising of economy, vegetable production played significant role in global trade, India has vast trading opportunity within and outside the country, and emerged as the world's second largest producer next to China. However, self-sufficiency in a real sense can only be realized if each person in India is assured by prescribed amount of vegetables, which is 300 g per person per day, including 120 g of green leaves, 90 g of tuber roots and 90 g of other vegetables. The consumption of vegetables per capita per day in India is far lower than the normal requirement, in order to address the problems of under-nutrition and malnutrition, the intake of vegetables must be significantly increased.

This book contains information for Teaching of Vegetable Science courses for B.Sc., M.Sc. and Ph.D. in Agriculture/Horticulture Vegetable Science, it provides fundamental knowledge, basic concepts and research advancements of Cucumber, Tomato, Okra, Snap Melon, Spine gourd and insights of production, breeding, diseases resistance and new aspects in vegetable crops, with details information on the different facets of vegetable crops. The book is focused on research records of vegetable scientists and researcher in different organizations working on vegetable crops, thus readers may inspire to develop their vegetable science knowledge in their future careers, undoubtedly, acquiring scientific knowledge would be rewarding for its consumers. To further facilitate the accessibility of the desired information in various areas covered in this collection, the book is divided into 9 sections.

Authors

Contents

Foreword *v*

Preface *vii*

List of Contributors *xi*

1. **Genetic Resources, Systematics and Taxonomy of Cucumber for Successful Utilization in the Breeding and Improvement** 1
J.C. Bommesh, R. Selvakumar, D.C. Manjunathagowda and T.G. Harishkumar

2. **Breeding for Downy Mildew Resistance in Cucurbits** 19
N. Shilpashree, M. Anjanappa, D.C. Manjunathagowda, M. Prakash and J.C. Bommesh

3. **Breeding Strategies for YVMV Resistance in Okra** 35
K.N. Kavya, D.C. Manjunathagowda, G. Prakash and R. Selvakumar

4. **Breeding of Vegetables for Nutritional Quality** 47
Abhilash Singh, Shubham Singh, Sandeep Yadav and Ramavath Ramesh Babu

5. **Disease Resistance Breeding in Tomato (*Solanum lycopersicum* L.)** 61
Ramavath Ramesh Babu, G. Indraja, Shubham Singh and Raman Selvakumar

6. **Speed Breeding in Vegetable Crops: A Smart Approach to Accelerate Cultivars Development** 105
Shubham Singh, R. Ramesh Babu, C. Anil Kumar and Sandeep Yadav

7. **Parthenocarpic Vegetables: Importance and Approaches** **123**

B. Arunkumar, G.C. Nagesh, D.C. Manjunathagowda and R. Selvakumar

8. **Snap Melon (*Cucumis melo* var. *momordica* Duthie and Fuller): A Minor Cucurbit** **145**

Jyothi Motapalukula, Mallesh Sanganamoni, Vittal L. Mangi and Ankit Panchbhaiya

9. **Spine Gourd (*Momordica dioica* Roxb. ex Willd.): An Indigenous Cucurbitacous Vegetable Crop** **159**

Mallesh Sanganamoni, Vittal L. Mangi, Ankit Panchbhaiya and Jyothi Motapalukula

Index **173**

List of Contributors

Anil Kumar, C.
Scientist, ICAR-National Rice Research Institute, Cuttack, Odisha

Anjanappa, M.
College of Horticulture, University of Horticultural Sciences, UHS Campus, Bengaluru–560065, Karnataka

Arunkumar, B.
Assistant Manager (Production), Central State Farm, National Seeds Corporation Ltd., Raichur–584143, Karnataka

Bommesh, J.C.
Department of Horticulture, Government of Karnataka, MS Building, Bengaluru, Karnataka

Harishkumar, T.G.
University of Agricultural Sciences, GKVK, Bangalore–560065, Karnataka

Indraja, G.
Dr. Y.S.R. Horticultural University, Andhra Pradesh

Kavya, K.N.
College of Horticulture, University of Horticultural Sciences, UHS Campus, Bengaluru–560065, Karnataka

Mangi, Vittal L.
College of Agriculture, Vijayapura, University of Agricultural Sciences, Dharwad–586101, Karnataka

Manjunathagowda, D.C.
Scientist, ICAR-Directorate of Onion and Garlic Research, Rajgurunagar, Pune-410505, Maharashtra

Motapalukula, Jyothi
College of Agriculture, VNMKV, Parbhani, Maharashtra, Karnataka

Nagesh, G.C.
Breeder-Research and Development (VC), Seed Works International Pvt. Ltd., Behind Sugar Factory, Chikkaballapura Dist–561208, Karnatka

Panchbhaiya, Ankit
Sectional Officer (Horticulture), DDA, Hauz Khas, New Delhi–110016

Prakash, G.
College of Horticulture, University of Horticultural Sciences, UHS Campus, Bengaluru–560065, Karnataka

Prakash, M.
Research and Development, Takkii and Chiathai India Research Private Limited, Bengaluru–561203, Karnataka

Ramesh Babu, R.
Division of Vegetable science, University of Horticultural Sciences, Bagalkot, Karnataka

Sanganamoni, Mallesh
College of Horticulture, Rajendranagar, SKLTSHU, Telangana–500030

Selvakumar, R.
Scientist, Centre for Protected Cultivation Technology, ICAR-Indian Agricultural Research Institute, Pusa Campus, New Delhi–110012

Shilpashree, N.
College of Horticulture, University of Horticultural Sciences, UHS Campus, Bengaluru–560065, Karnataka

Singh, Abhilash
Department of Vegetable Science, G.B Pant University of Agriculture and Technology Pantnagar, Uttarkhand

Singh, Shubham
Division of Vegetable Science, ICAR- Indian Agricultural Research Institute, Pusa Campus, New Delhi–110012

Yadav, Sandeep
G.B. Pant University of Agriculture and Technology, Pantnagar, Uttarakhand

Chapter 1

Genetic Resources, Systematics and Taxonomy of Cucumber for Successful Utilization in the Breeding and Improvement

J.C. Bommesh, R. Selvakumar, D.C. Manjunathagowda and T.G. Harishkumar

ABSTRACT

"The family Curcurbitaceae consists of 118 genera and 825 species in two sub-families, Zanonioideae and the Cucurbitoideae. The Cucurbitoideae houses eight tribes, one of which Melothrieae includes the genus Cucumis, Cucumis is partitioned into two groups Cucumis (2n=2x= 14 and 24) and Melo (2n=2x=24) that contain five cross-sterile species groups (Jeffrey 1980). The subgenus Cucumis comprises three or four Sino-Himalayan species, including Cucumis sativus (2n=2x=14) and Cucumis hystrix Chakr. (2n=2x= 24). C. sativus houses several botanical varieties including var. sativus, the cultivated cucumber (C. sativus var. sativus), and the wild, free-living var. hardwickii (R.) Alef. (C. sativus var. hardwickii). Wild African Cucumis species (2n = 2x = 24) are cross incompatible with cucumber and melon, which are themselves incompatible. The wild, free living C. hystrixis only sparingly fertile with cucumber. It is found only in the Yunnan Province of Southern China, and has unique genetic attributes that make its taxonomic determination complex".

Keywords: *Cucumber, Curcurbitaceae, Cucumis, Genera.*

Introduction

Cucumber (*Cucumis sativus* L.) is a member of the economically important family Cucurbitaceae with chromosome number of 2n = 14. It is grown for edible tender fruits as dessert, pickles and cooked vegetables for its cooling effect. It is useful for preventing constipation and beneficial for people suffering from jaundice and allied diseases. Cucumber seed contain oil, which is useful for brain development and body smoothness. Hence, it is being used for Ayurvedic preparation. The world area is 1.972 million hectares with production of 62.431 million metric tonnes and in India it is cultivated in an area of 0.24 lakh hectares with an annual production of 1.52 lakh metric tonnes (NHB, 2015).

Taxonomic Position

Kingdom	:	Plantae
Division	:	Magnoliophyta
Class	:	Magnoliopsida
Order	:	Cucurbitales
Family	:	Cucurbitaceae
Genus	:	*Cucumis*
Species	:	*Sativus*

Diagnostic Characters to the Genera *Cucumis*

Cucumis

Fruits are spiny and non netted.

Fruits are glabrous, pubescent, smooth, netted, mostly with sutures. Ex- *Cucumis melo.*

Leaves are deeply lobed, small fruits prickly, lemon yellow when mature. Ex- *C. anguria*

Leaves are shallow lobed, large and ablong fruits. Ex- *C. sativus*

The genus *Cucumis* consists of 32 species, which are of annual and perennial nature, which are classified as two distinct groups of geographic origin and chromosome number *viz.* African group ($2n$ = 24) and Asiatic group ($2n$ = 14) (Kirkbride 1993). The African group contain musk melon (*C. melo* L.), and the Asiatic group consist of cucumber (*C. Sativus* L.) and its probable ancestor *C. sativus* L. var. *hardwickii* (Royle) Synonymous *C. hardwickii* (Perltreves and Galun 1985).

The isozymes, cpDNA and restriction fragment length polymorphism markers supported the distinction between melon and cucumber (Perl-Treves and Galun

1985, Perl-Treves *et al.*, 1985). There are two theories that attempt to explain the existence of a haploid chromosome number $n = 7$ and a close relative with a haploid chromosome number $n = 12$. The haploid chromosome number may have been increased ($n= 12$) by fragmentation or reduced ($n = 7$) by fusion.Fusion is supported by the increased amount of heterochromatin content in *C. sativus* (Ramachandran and Narayan, 1985) and by the presence of chromosomal alterations (karyotypic) in *Cucumis* speciation (Sing and Roy, 1974). Comparative genomics between *C. melo* and *C. sativus* may clarify the phylogeny of these species (Danin-Poleg *et al.*, 2001). Cucumber originated in India about 3,000 years ago and was soon cultivated in the South and East of the Himalayas, forming the Asiatic group (Kroon *et al.*, 1979; Ramachandran and Narayan, 1985). From India, cucumber was brought to Greece and Italy and later to China. Records confirmed cucumber cultivation in Francein the 9th century, England in the 14th century and in North America by the mid-16th century (Swiader *et al.*, 1992).

Botany

The key characters of the genus are fruit fleshy, many seeded *pepo*, flowers solitary, flowers lemon–yellow to deep orange, leaves deeply or shallowly lobed not pinnatified, and corolla rotate, deeply 5 parted and small.

Etymology

Cucumis in Latin means cucumber, the word cucumber was derived from Greek word *kykyon*. The epithet *sativus* is a Latin word, means 'that is sown', referring to the common agricultural use of the species

Habitat

Herbaceous vine with pubescent stems and unbranched tendrils, it will grow in well-drained (sandy) soils rich in organic matter and slightly alkaline. It prefers full sun exposure in warm and humid climates and it is not frost resistant.

Description of Cucumber (*C. sativus*)

Plant Morphology

Cucumbers grow up to 2m to 5m tall, long leaves up to 30cm, leaves are alternate and simple, with 3-7 palmate lobes and serrated margins, yellow 5-merous flowers bearing either female or male organs hairless cylindrical fruits are warty, yellow to green, and up to 50 cm long.

Vegetative Plant

C. sativus is a morphologically variable annual herbaceous climber. The stems are prostrate, angular, and covered in white pubescence. Stipules are absent, and the plant bears unbranched axillary tendrils up to 30cm long 10-16cm long petioles. The pubescent leaves are alternately arranged on 10-16 cm long petioles, simple, basally cordate, and apically acute with 3-7 palmate lobes. The palmately-veined leaves are nearly orbicular, 7-20 cm long and broad.

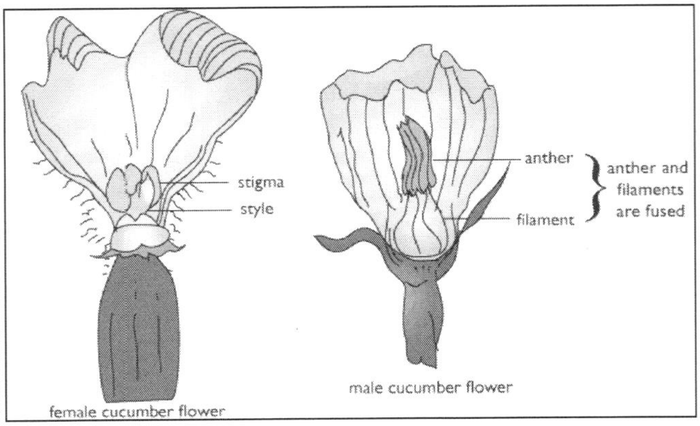

Flower

C. sativus is monoecious (separate male and female flowers on the same plant) and its axillary flowers are actinomorphic and rarely bisexual. Both staminate and pistillate flowers have a pubescent 5-parted calyx composed of 0.5-1cm long white pubescent sepals. The sepals are long, narrow, and acute; on pistillate flowers, the calyx is fused to the ovary and forms a hypanthium. The corolla is approximately 2cm long, yellow, fused less than half of its length, campanulate, 5-parted with oblong to lanceolate lobes. Staminate flowers are solitary or 3-7 on pubescent pedicels (0.5-2cm long), bearing 3 stamens, of which two have 2-celled anthers (0.3-0.4cm long) and one a 1-celled anther. Pistillate flowers are solitary or paired on pedicels shorter than the staminate (<0.5cm long) before fruit development, whereas during fruit development the pedicels can elongate up to 5cm long. The pistilate flowers have 3 staminodes, a pubescent, ellipsoid, unilocular ovary 2-5cm long, with a short style and 3 stigmas.

Floral Biology

The whole developmental process from initial bud stage to the stage when the flower is detached from the pedicel was divided into 8 stages (Stage 1 to 8). Opening and closing of the male flowers are mainly influenced by sun rise and sun set, which is by light and time of the day. Anthesis takes place around 5.30 – 7.00 hr. dehiscence occurs around 4.30 – 5.00 hr. pollen fertility is upto 14 hrs.

Pollination

Cucumber is pollinated by bees that harvest the nectar produced by the flowers; honey bees are more effective than wild bees. The pollen is sticky, which makes wind pollination improbable.

Fruit

The fruit is an indehiscent cylindrical berry with many seeds. Cucumbers are glabrous, and can be smooth or warty, yellow or green, ranging from 5-100 cm long and weigh 50g to 4kg. Each plant produces up to 25 fruits.

Seed

The white seeds range from 0.5-1.8 cm in length, and are emarginate and elliptical with pointed ends. A gram of seed includes 33-40 seeds.

Origin and Distribution

Cucumber is thought to be one of the oldest vegetables cultivated by man with historical records of 5,000 years ago (Wehner and Guner, 2004). Its diversity occours between the Himalayas and Bay of Bengal area. It is originated at foothills of the Himalayan Mountain, where its two botanical varieties were discovered, domesticated cucumber *Cucumis sativus.* var. *sativus* and wild progenitor of cucumber *Cucumis sativus* var. *hardwickii* (Lv *et al.*, 2012). Among all species, *C. sativus* and to a limited extent *C.callosus* (Rottle.) Cogn. are cultivated, while *C. hardwickii, C. hystrix, C. prophetarum* L. and *C. setosus* Cogn. are in wild forms. Sebastian *et al.* (2010) reported that cucumber is Asian origin and Southeast Asian *Cucumis hystrix* is the closest relative of cucumber by using DNA sequences from plastid and nuclear markers, for some 100 *Cucumis* accessions from Africa, Australia, and Asia.

"Cucumber was brought to Greece and Italy by the Romans (2nd century BC; Mesopotamia), and it appeared in France in the 9th century, in England in the 14th century, and in North America by the mid-16th century. Cucumber's dissemination westward from India is indicated by the profusion of ancient names that describe it. The english word "cucumber" comes from the latin name *Cucumis*. Likewise, the Bohemian *agyrka*, German *gurke*, Greek *Aggouria*, and European *gherkin* trace back to an ancient Aryan word. *Cucumis sativus* var. *hardwickii* is a wild relative of *C. sativus* var. *sativus* that grows in the foothills of the Himalayan Mountains and is used by native peoples of Northern India as a laxative (Deakin*et al.*, 1971)".

"The genetic variation in *C. sativus* var. *sativus* accessions from India and China (the secondary centre of diversity for cucumber) has been assessed by protein and DNA marker analyses (Staub *et al.*, 1997; 1999). Data indicate that variation exists among accessions from different Indian states, the differences between Indian and Chinese accessions are distinct, and that Indian and Chinese accessions themselves are distinctly different from other *C. sativus* var. *sativus* genotypes throughout the world. Genetic differences exist among cultivars grown in the same growing region [*e.g.*, Shanghai and the Hunan region (Southern China) and Jiang Su, and Anhui (Northern China)]. These facts suggest that cultivar differences among Indian and Chinese cultivars can exist over a relatively limited geographical range".

Classification of Diversity

Botanically classified into 4 sub species

1. *C. sativus* ssp. *rigidus* – English long fruited of Eastern Asiatic type
2. *C. sativus* ssp. *agrestis* – Soviet Far East type
3. *C. sativus* ssp. *gracillis*
4. C. *sativus* ssp. *sphaerocarpus* (Syn to *C. spaerocarpin*)

Table 1.1. Diagnostic Feature and Distribution of Wild Species of *Cucumis*

Sl. No.	Species Name and Chr. No	Distribution	Diagnostic Features	Fruiting Season
1	*Cucumis callosus* 2n =24, Bislumbha	World: S.E. Asia, Africa, Australia. India: Andhra Pradesh, Gujarat, Haryana, Uttar Pradesh, Tamil Nadu, MH, Orissa and WB	Perennial creeping herb with woody rootstock, Fruit is indehiscent, ovate to globose, green with ten yellowish green to white continuous longitudinal stripes, turns yellow and glabrous at maturity, bitter. Seed is ovate to lance-ovate, beaked, funiculus persistent.	Aug-December
2	*Cucumis dispaceus* 2n =24, Hedgehog gourd	West Indies, Africa, North and South America, Saudi Arabia **India:** Tamil Nadu and Karnataka	Annual climbing herb. Leaves cordate-orbicular. Fruit indehiscent, oblong, green when young pale yellow at maturity, densely aculeate, bitter. Both ends of seed is acute, ovate to lance-ovate, funiculus caducous, non-beaked, grey-white.	July-Jan
3	*Cucumis hystrix* 2n = 24, Arphagma (Mizoram)	**India**: Mizoram	Annual climbing herb. Hairs antrose. Leaves pentangular, middle lobe distinctly elevated and shallow incised at base. Fruit indehiscent, ovate, tapering at both ends, beaked, aculeate; aculi obtuse, green. Seed apex rounded, base acute, ovate to lance-ovate, funiculus caducous, non-beaked, gery-white.	July-December
4	*Cucumis hystrix* 2n = 24	China, Myanmar, Thailand **India:** Arunachal Pradesh, Meghalaya, Mizoram and Nagaland.	Annual climbing herb. Hairs retrose to perpendicular to surface. Leaves pentangular. Fruit indehiscent, ovate, tapering at both ends, beaked, aculeate; aculi acute, green. Seed apex rounded, base acute, ovate to lance-ovate, funiculus caducous, non-beaked, grey-white.	August to December
5	*Cucumis indicus* Meti, Tausal 2n =20	Endemic to Northern-Western Ghats of India	Annual climber. Leaves angular. Fruit indehiscent, elongated, fusiform, rostrate, pale green with 10 dark green longitudinal stripes, turn pale green to white at maturity, glabrous. Seed apex rounded, base acute, ovate to lance-ovate, funiculus caducous, non-beaked, grey-white.	August to November
6	*Cucumis javanicus*	-	Annual climber herb. Leaves pentangular. Fruit ellipsoid, pale green, turning dark red at maturity. Seed is elliptic to obovate, apex rounded, beaked, marginated, smooth, grey-white.	August to November
7	*Cucumis leiospermus*	Sri Lanka **India**: Tamil Nadu, Kerala	Annual climbing herb. Leaves pentangular. Fruit is oval-globose, indehiscent, pale green with 10 dark green longitudinal stripes, turning dark red at maturity, 16-17 seeded. Seed is ovate to lance-ovate, funiculus caducous, apex rounded, short beaked, marginated, smooth, grey-white.	August to December

Sl. No.	Species Name and Chr. No	Distribution	Diagnostic Features	Fruiting Season
8	*Cucumis maderaspatanus* 2n = 24 Chirati(Marathi), Mukkapira (Malayalam)	Wide spread. Africa, SW and SE Asia, Australia. **India**: In all the states	Annual climbing herb. Leaves pentangular. Fruit circular, pale green with 7 -10 dark green longitudinal stripes, turning dark red at maturity, 3-20 seeded. Seed is elliptic to obovate, apex rounded, beaked, marginated, smooth when fresh, scorbiculate when dry, grey-white.	August to November
9	*Cucumis melo var. agrestis* 2n =24 Kachoori, Tkmak, Shinde (Marathi), Simmatikkai (Tamil)	Worldwide. India: All states	Annual prostrate, herb. Leaves shallowly 3-5 lobed. Fruits are indehiscent, ovate – oblong, mottle dark green with 10 pale green continuous longitudinal stripes, turns pale yellow at maturity, mostly bitter, rarely not bitter. Seeds are ovate to lance-ovate, beaked, funiculus persistent, grey-white.	June to December
10	*Cucumis melo* 2n = 24	World wide India: All states	Annual prostrate, herb. Leaves shallowly 3-5 lobed. Fruits are indehiscent or dehiscent, oval, globose or elliptic, bicoloured or monocoloured, green, turns yellow or orange at maturity, edible or inedible, bitter or not bitter in taste. Seeds are ovate to lance-ovate, beaked, funiculous persistent, grey-white.	June to February
11	*Cucumis muriculatus*	Burma	Annual climbing hairy herb. Leaves pentangular. Fruits are indehiscent, oblong-round, tapering at both ends, beaked, densely echinate-muriculate, green. Seeds apex rounded, base acute, ovate to lance-ovate, funiculus caducous, non-beaked, grey-white.	
12	*Cucumis prophetarum* Khat-Kachario (Rajasthan) 2n = 24	Ethiopia, Kenya, Somalia, Egypt, Uganda. Middle East, Pakistan **India**: Gujarath, Karnataka, Kerala, Maharashtra, Rajasthan and Tamil Nadu	Perennial creeping herb, with woody root stock. Leaves deeply 3-5 lobed. Fruits are oval to globose, green with ten white continuous longitudinal stripes, green when young turns yellow at maturity, bitter, aculeate. Seeds both ends acute, ovate to lance-ovate, funiculus caducous, non-beaked, grey-white.	August to January
13	*Cucumis ritchiei* 2n = 24 Ghugarya (Marathi)	**India:** Gujarath, Maharashtra, Karnataka. Endemic to Northern-Western Ghats	Annual climbing herb. Leaves pentangular. Fruits are globular, indehiscent, green with ten white longitudinal stripes, turns red or black at maturity, glabrous. Seeds are ovate to rectangular, beaked, base rounded with raised part of margin, turgid, funiculus caducous, grey-white, margin distinct, three chambered; two lateral chambers empty, middle with cotyledons.	August to November

Sl. No.	*Species Name and Chr. No*	*Distribution*	*Diagnostic Features*	*Fruiting Season*
14	*Cucumis sativus* 2n = 14	World wide India: All states	Annual climbing, herb. Leaves pentangular. Fruits are indehiscent, oblong, rounded or cylindrical, monolocular or bilocular, turns pale yellow, yellow orange or brown at maturity, eduble or inedible, bitter or non bitter in taste. Seeds are ovate to lance-ovate, beaked, funiculus persistent, grey-white.	Through-out the year
15	*Cucumis sativus* forma *hard-wickii* Jangali Kakadi (Marathi), Wild cucumber (English) 2n = 14	World wide India: All states	Annual climbing, herb. Leaves pentangular. Fruits are indehiscent, oblong to rounded, green with ten longitudinal stripes, rarely white, green when young turns pale yellow to brown at maturity, bitter. Seeds are ovate to lance-ovate, beaked, funiculus persistent, grey-white.	June to December
16	*Cucumis setosus* Meki, Mehaki, Mekunya (Marathi) 2n = 24	Endemic to Northern-Western Ghats. Edible	Annual herb. Leaves pentangular. Fruits are indehiscent, pulp granular, oblong, yellowish green with ten dark green longitudinal stripes, turns pale green to whitish at maturity, setose hairy, Seeds are ovate to lance-ovate, apex rounded, base acute, non-beaked, funiculus caducous, grey-white.	July to December
17	*Cucumis silentvalleyi* 2n = 24	Kerala and Tamil Nadu. Endemic to Southern-Western Ghats	Annual climber. Leaves angular. Fruits are dehiscent, oblong, fusiform, rostrate, green to pale green in colour with whitish spotted ten longitudinal stripes, turn pale green to white at maturity, retrose hairy, hairs bulbous based. Seeds are ovate to lance-ovate, apex rounded, base acute, non-beaked, funiculus caducous.	August to November

Classification of Cucumber based on cultivars groups (Rubatzky, Yamaguchi 1997; George 1999)

1. Field cucumbers, with prominent white or black spines
2. English cucumbers – greenhouse or forcing type of parthenocarpic cucumbers with seedless fruits.
3. Sikkim cucumber - originated in India, with reddish brown skins of fruits;
4. Oriental-type cucumber- popular in Japan and China, thin skinned, usually long and slender with conside-rable warts and spines.
5. Smoot-skinned Beit Alpha types- most popular in Middle East and north Africa, with light coloured, thin skinned fruit.
6. Lemon cucumber - produces a round, creamy yellow skin fruit with large five-carpel seed cavity. Its sex expression is andromonoecious.
7. Gherkins - very small, immature cucumber fruits.

Cytotaxonomy (Cytology)

The genus *Cucumis* comprises about 30 species distributed over two distinct geographic areas (Jeffrey, 1980; Chen and Adelberg, 2000)

1. **South-East Himalaya:** An important origin of the Asiatic group with chromosome of X =7 (Cucumber belongs).
2. **African group:** Group consisting species with *basic chromosome* X=12 (Muskmelon). The evolution of chromosomes in *cucumis* species is not clear. Whitaker (1933) assumed that species with 2n=24 chromosomes arisen from species with 2n=14 chromosomes by fragmentation but later Trivedi and Roy (1970) reported that by fusion of chromosomes interminal centromere species with 14 chromosome have arisen from species with 2n=24 chromosomes. Karyologically, the varieties have 2n = 14 chromosomes, and possess three pairs of chromosomes with secondary constrictions besides, other 14 chromosome species C. *callosus* Rottle syn *C. trigonus* Roxb. Cytological investigations have long been recognized to bring about a clear understanding of phylogeny and evolution of crop plants, but such studies have not been rewarding in the genus *Cucumis* due to small size of mitotic chromosomes and their poor stainability. In an intensive cytogenetical attempt with C″ banding and pachytene analysis, Ramachandran and Seshadri (1986) compared the genomes of C. *sativus* and C. *melo* (Table 1.2). Differences of basic karyotype also influenced to the physical mapping of a tandem repeat sequence detected by fluorescence *in-situ* hybridization (FISH) method (Hoshi *et al.*, 1999, Chen *et al.*, 1999, Koo *et al.*, 2002). Ribosomal RNA gene (rDNA) repeats were detected in several cucumber cultivars. However, the descriptions of positions, numbers and sizes of the rDNA were not connected to those of the previous reports. In this study, DAPI-bands and the positions of 5S and 45S rDNA sites in the three cultivars of cucumbers from Poland, Japan and China were compared to detect their chromosome relationships (Tagashira *et al.*, 2006).

Table 1.2. Karyotypic Characters of Mitotic Chromosomes of *C. sativus* and *C. melo*

Karyotype Characters	*C. sativus*	*C. melo*
Diploid chromosome number (2n)	14	24
Type of chromosomes		
Median	2	8
Sub-median	4	2
Sub-terminal	1	1
Satellite pairs	3	2
Range in chromosome length (*μm*)	1.25-2.72	1. 06-1.88
Average chromosome length (*μm*)	1.96	1.42
Largest/shortest chromosome ratio category of karyotype of Stebbins (1958)	2.18:1 Lb	1.13:12a

Source: Ramachandran and Seshadri (1986).

Species Forming the Asiatic Group

C. sativus var. *hardwickii* (2n= 2x = 14) is a progenitor of cultivated cucumber, it's a small bitter cucumber with scattered, short, sharp, stiff and often evanescent spines. It possesses a multiple fruiting and branching habit not present in the cultivated cucumber (Horst and Lower, 1978). It possesses useful characters such as prolific fruit bearing, high number of laterals. Polymorphism level in *C. sativus* var. *hardwickii* is higher than in *C. sativus* var. *sativus*. This species is widely distributed in foothills of north-western Himalayas (parts of Uttaranchal and Himachal Pradesh) followed by fair distribution in Western Ghats (parts of Maharashtra, Goa, Karnatka, Kerala and Tamil Nadu) and sporadic distribution in Eastern Ghats and parts of Rajasthan (Mt. Abu), at elevations from 800 to 1700 MSL. Its existence in parts of north-eastern region of India is yet to be explored (Bisht *et al.*, 2004).

Cucumis hystrix- wild *Cucumis* ssp. indigenous to Yunnan province of China. It has been referred to in several taxonomic and systematic studies (Dane, 1991) and has been classified as a species in the subgenus *Cucumis* based on its morphology. However, nothing was known, except for its morphology, until it was rediscovered in natural wild conditions in 1989 and subjected to subsequent genetic characterization by Chen *et al.* (2002). This is the first Asian *Cucumi* ssp. described with chromosome number 2n=24. This finding challenges the basic chromosome number theory that African *Cucumis* have x=12, and that Asian *Cucumis* have x=7, which has governed the understanding of systematics and phylogenetic in *Cucumis* for decades. Although *C. hystrix* possess the same number of chromosomes as *C. melo* yet its fruit has a typical cucumber taste and flavour. Moreover, *C. hystrix* has been found to be closer to *C. sativus* than to *C. melo* on the basis of isozyme patterns (Chan *et al.*, 1995), SSR and RAPD marker analysis (Zhuang *et al.*, 2004) and CCSSR marker sequence variation. Besides China, this species is also distributed in India (Assam), Burma and Thailand.

Evolution of Species *Cucumis*

"The biosystematics and phylogeny of *Cucumis* species based on morphology, crossability, and protein analysis (Deakin*et al.*, 1971; Staub *et al.*, 1987; 1992) has led to an understanding of species relationships that have been largely confirmed by nuclear DNA analysis (Jobst *et al.*, 1998). Most recently, Garcia Mas *et al.* (2004) defined phylogenetic relationships among *Cucumis* species using ribosomal internal transcribed spacer sequences and microsatellite markers. Although their data did not agree with some of the previously described genetic relationships obtained using isozyme and restriction fragment markers, their description of a clear separation between *C. sativus* and the rest of the *Cucumis* species supported these earlier studies. The center of origin for *Cucumis* species is likely Africa for the wild species. However, initial sites of domestication for melon and cucumber are probably the Middle East and Southern Asia, respectively, where genes from exotic sources have contributed extensively to plant improvement (Dane *et al.*, 1980)".

Possible Evolution of *Cucumis* Species

C. melo L. var. *reticulatus*
(Melon are Netted and Sweet taste)

↑

C. melo L. var. *cantalupensis*
(Melons are pubescent, lorky, warty, ridges and folds, blend, slightly sweet and sour)

↑

C. melo L. var. *acidulus*
(Melons are pubescent, smooth and sour)

↑

C. melo L. var. *callosus*
(Melons are pubescent and smooth, bitter types and sweet types)

↑

C. sativus L. (few or no soft spines and blend)	*C. sagittatus* (Melons are hairy and bitter)	*C. ficifolious* (soft spines and bitter)
↑	↑	↑
C. hardwickii (spiny and bitter) 2n = 14	*C. anguria* (Spiny and very bitter)	*C. heptadactylus* (Thick spiny and bitter) 2n = 24
	↑	
	C. metuliferus (large stout spines, very bitter) 2n = 24	

Evolution of Synthetic Amphidiploid Species in *Cucumis*

Synthetic species in *Cucumis* was developed by cross between *C. hystrix* (2n=24) and *C. sativus* (2n=14) was made that resulted in the creation of a sterile F_1 hybrid (2n= 19) by embryo rescue (Chen *et al.*, 1997b). After chromosome doubling and fertility selection, the primary amphidiploid (2n=38) produced fertile flowers and set fruit with viable seeds. These amphidiploids are in fact cross-compatible with *C. sativus* such that backcrossing to *C. sativus* allows for the introgression of genes from *C. hystrix* (*e.g.* nematode resistance) at relatively high ploidy levels (2n=38) (Chen *et al.*, 2002). Chen *et al.* (2003a) have recovered allotriploid plants (2n=3x=26) from the culture of embryos derived from a *C. hytivus* x *C. sativus* mating. These plants possessed characteristics common to both their *C. hytivus* and *C. sativus* parents. These plants were highly male sterile, their progeny were more parthenocarpic (84.8 per cent) than their parents (26.7 per cent and 45.2 per cent). In China and perhaps other countries such allotriploid plants have potential for its pickling industry. However, frequent subculture of the plant may reduce its proliferating capacity and increase its genetic variation (*e.g.* somaclonal variation). The allotriploid plants obtained from micropropagation are uniform and possess unique characteristics that include sequential fruit set and tolerance for growth under low temperature and light (Chen *et al.*, 2003b). These allotriploid plants can produce unreduced pollen and when used in conjunction with genetic engineering technologies such as alien addition, substitution and translocation lines in repetitive backcrossing strategies can allow for the introgression of foreign genes into commercial cultivars (Barthes and Ricroch, 2001).

Phylogenetic Affinities among *Cucumis* Species

Phylogenetic relationships among *Cucumis* species have been studied on the basis of isozyme analysis (Perl-Treves, 1985), plastome DNA restriction analysis, ITS data, SSR marker sequence variation and RAPD marker analysis (Jobst, *et al.*, 1998). Although the number of groups varied with each study, all of these reports agreed upon two general features of *Cucumis* phylogeny that are also recognized by more classical approaches:

a) The clear separation between C. *sativus* and the rest of the species in the genus
b) The separation of *C. metuliferus, C. melo* and the group of wild African species as three separate entities within the x=12 species.

Garcia Mas *et al.* (2004) have suggested a further division of the group of wild species into two clades first would include *C. sagittatus* (synonymous *C. dinteri*), with the possible addition of *C. globosus* and *C. humifructus* and second containing the rest of the wild species, would be further sub divided into two sub clades; one including *C. prophetarum, C. ficifolius* and *C. membranifolius* and the other containing the remaining species. Moreover, Chen *et al.* (2005) have proposed the creation of a provisional *Cucumis* species complex to include a subgenus *Cucumis* containing three series; *hystrix, hytivus* and *sativus* but this hypothesis would require rigorous testing and would be the initial step in the precise taxonomic placement of *C. hystrix*, *C. hytivus* and *C. satvius*.

The first comprehensive crossability analysis of the genus *Cucumis* was published by Deakin *et al.* (1971), who have divided 16 *Cucumis* species into four cross-sterile groups:

1. **Group I** included 9 species *viz. C. anguria, C. anguria* var. *longipes, C. dipsaceus, C. africanus, C. leptodermis, C. myriocarpus, C. prophetarum, C. zeyheri, C. ficifolius* and *C. heptadactylus.*
2. **Group II** included only one species *i.e. C. metuliferus* which was not closely related to any other species
3. **Group III** comprised *C. sativus* and *C. hardwickii*
4. **Group IV** consisted of *C. melo, C. humifructus, C. sagittatus* and *C. dinteri.*

Based on pollen tube behavior, *C. africanus* and *C. melo* appear to be the most promising male parents for crossing with *C. sativus,* but special pollination techniques and advanced embryo culture methods are necessary to overcome interspecific barriers in several crosses (Kho *et al.*, 1980). Other more successful interspecific hybridization studies of *C. melo* and *C. sativus* with wild species are presented in Tables 1.3 and 1.4. However, in practice, most of these results were not repeatable and did not result in fertile hybrids.

The cross between *C. sativus* and *C. hystrix* was the first repeatable cross between a cultivated *Cucumis* species and a wild relative (Chen *et al.*, 1997b) and represented a breakthrough in interspecific hybridization in *Cucumis.* Firstly, if *C. hystrix* and *C. melo* are cross-compatible and if the F derived from either interspecific hybridization can be made fertile through crossing and/or chromosome doubling, then *C. hystrix* could act as a bridge species between *C. melo* and *C. sativus*. Secondly, the fertile synthetic amphidiploid species *Cucumis hytivus* may be useful as a bridge species for the introgression of desirable genes from *C. hystrix* into *C. sativus* by either conventional intercrossing or transformation (Chen *et al.*, 2003a). Thirdly, if *C. hystrix* and *C. sativus* var. *hardwickii* are cross compatible then opportunities will exist to introgress economically important genes from *C. hystrix* and its derivatives into advanced *C. sativus* var. *hardwickii* derived lines for cucumber improvement (Zhuang *et al.*, 2004).

Table 1.3. Crossing Attempts with *C. sativus*

Cross	*Results*	*References*
C. zeyheri x *C. sativus*	Fruit with unviable seeds	Custers and Den Nijs (1986)
C. sativus x *C. metuliferus*	Embryos only	Franken *et al.* (1988)
C. sativus x *C. hystrix*	Sterile plants (2x and 4x)	Chen *et al.* (1997)
C. hystrix x *C. sativus*	Fertile plants	Chen *et al.* (1998)

Source: Chen and Adelberg, 2000.

Taxonomy of the Family Cucurbitaceae

The genus *Cucumis* belongs to the family Cucurbitaceae, order Cucurbitales. According to specific morphological features of tendrils, pollen grains and ovules, there are clear relation of this taxon with the order Passiflorales. Based on latest knowledge of cytology, cytogenetics, phytochemistry and molecular genetics. The family is subdivided into two subfamilies. The subfamily Zanonioideae comprises species with a low economic impact. All economically important species are included in the subfamily Cucurbitoideae (Table 1.1).

Within Cucurbitaceae the tribe Melothrieae with the genus *Cucumis* are considered the most important ones under climatic conditions of the Middle Europe. The genus *Cucumis* is represented by 32 species. Besides cucumber (*Cucumis sativus* L.) and melon (*C. melo* L.), the species *C. anguria* (West Indian Gherkin) and *C. metuliferus* (African Horned Cucumber) are commercially explored in several areas as well. Other wild species originating mostly from arid and/or semi-arid regions of Africa, respectively, are cultivated as ornamental plants (*C. dipsaceus* "Hedgehog Gourd", *C. myriocarpus* "Gooseberry Gourd")

Table 1.4. Source of Resistance to Biotic Stress, Abiotic Stress and Quality Characters

Sl.No.	Plant Introduction	
1	PI 183056	India; large root size
2	PI 183967	India; multiple lateral branching, sequential fruiting, nematode resistance
3	PI 197087	India; downy mildew resistance
4	200815	Myanmar; powdery mildew and gummy stem blight resistance
5	PI 209065	U.S.; high yield
6	PI 212233	Japan; powdery mildew resistance
7	PI 220860	South Korea; gynoecy
8	PIs 418962	China; multiple disease resistances
9	EC320556	CMV, scab and powdery mildew resistance
10	EC329300	Multiple disease resistance, gynoecious line
11	EC388737-39	Gynoecious lines
12	EC398030	Early determinate type
13	EC398030	Early, cluster bearing, determinate type
14	EC398966-67	Angular leaf spot resistance
15	EC398968-70	Anthracnose resistance
16	EC398971-73	Fruit rot resistance
17	EC398974-90	Downy mildew resistance
18	EC398991, 399007	Leaf spot resistance

Table 1.5. Sources/Donors Identified for Various Traits in Cucumber

Ttraits	*Donors*
Extended shelf-life	IC203838,IC203839
Early and determinate	EC398030
Gynoecious line	EC382739
High yield	IC203838, EC237658,VJ/98-176, VJ/98-151
Anthracnose resistance	PI197087, Poinsett
Downy mildew resistance	PI197087
Powdery mildew resistance	Poinsett, Yomaki, PI79376
Cucumber scab resistance	Wisconsin SMR9
Angular leaf spot resistance	Poinsett, MSU9402, PI169400
Bacterial wilt resistance	PI200815,PI200818
Cucumber green mottle mosaic resistance	*Cucumis anguria, C. africanus, C. ficifolius*
White fly resistance	*Cucumis asper, C. dinteri, C. sagittatus*

REFERENCES

Barthes L, Ricroch A (2001) Interspecific chromosomal rearrangements in monosomic addition lines of *Allium*. Genome 44: 929-935.

Bisht IS, Bhat KV, Tanwar SPS, Bhandari DC, Joshi K, Sharma AK (2004) Distribution and genetic diversity of *Cucumis sativus* var. *hardwickii* (Royle) Alef in India. J Hort Sci Biotech 79: 783-791.

Chen JF, Staub JE, Qian C, Jiang J, Luo X, Zhuang F (2003b) Reproduction and cytogenetic characterization of interspecific hybrids derived from *Cucumis hystrix* Chakr. x *Cucumis sativus* L. Theor Appl Genet 106: 688-695.

Chen JF, Adelberg J (2000) Interspecific hybridization in *Cucumis*-progress, problem and perspectives. Hort Sci 35: 11-15.

Chen JF, Zhuang FY, Liu XA, Qian CT (2004a) Reciprocal differences of morphological and DNA characters in interspecific hybridization in *Cucumis*. Can J Bot 82: 16-21.

Chen JF, Isshiki S, Tashiro Y, Miyazaki S (1995) Studies on a wild cucumber from China (*Cucumis hystrix* Chakr.) 1. Genetic distances between *C. hystrix* and two cultivated *Cucumis* species (C. *sativus* L. and C. *melo* L.) based on isozyme analysis. J Jap Soc Hort Sci 64 (2): 264-265.

Chen JF, Luo XD, Staub JE, Qian CT, Zhuang FY, Ren G (2003a) An allotriploid derived from an amphidiploid x diploid mating in *Cucumis* 1: Production, micropropagation and verification. Euphytica 131: 235-241.

Chen JF, Staub JE, Adelberg JW, Lewis S, Kunkle B (2002) Synthesis and preliminary characterization of a new species (amphidiploid) in *Cucumis*. Euphytica 123: 315-322.

Chen JF, Staub JE, Tashiro Y, Isshiki S, Miyazaki S (1997b) Successful interspecific hybridization between *C. sativus* L. and *Cucumis hystrix* Chakr. Euphytica 96: 413-419.

Chen JF, Zhuang FY, Staub JE, Qian CT (2005) Genetic relationships among several wild and cultivated Asian and African *Cucumis* species. *Gen Res Crop Evol* (In press).

Dane F (1991) Cytogenetics of the genus *Cucumis*. In: *Chromosome Engineering in Plants : Genetics, Breeding. Evolution.* Part B. (Eds. Tsuchiya, T and Gupta PK) Elsevier, Amsterdam, pp 201-214.

Dane F, Denna DW, Tsuchiya T (1980) Evolutionary studies of wild species in the genus *Cucumis*. Z. Planzenzüchtg. 85: 89-109

Deakin JR, Bohan GW, Whitaker TW (1971) Interspecific hybridization in *Cucumis*. Economics Botany 25: 195-211.

Deakin JR., Bohn GW, Whitaker TW (1971) Interspecific hybridization in *Cucumis*. Econ. Bot. 25: 195-211.

Garcia-Mas J, Monfonte AJ, Arus P (2004) Phylogenetic relationships among *Cucumis* species based on the ribosomal internal transcribed spacer sequence and microsatellite markers. Plant Syst Evol 248: 191-203.

Horst EK, Lower LR (1978) *Cucumis hardwickii*: A source of germplasm for the cucumber breeder. Cucurbit Genet Coop Report 1-5.

Hoshi Y, Pl[1]der W, Malepszy S (1999) Physical mapping of 45S rDNA gene loci in the cucumber (*Cucumis sativus* L.) using fluorescence *in situ* hybridization. Caryologia 49-57.

Jeffrey C (1980) A review of the cucurbitaceae. Bot J Linn Soc 81: 233-247.

Jobst J, King K, Hemleben V (1998) Molecular evolution of the internal transcribed spacers (ITS1 and 1TS2) and phylogenetic relationships among species of the family cucurbitaceae. Mol Phylogenet Evol 9: 204-219.

Kerje T, Grum M (2000) The origin of melon, *Cucumis melo*: a review of the literature. Acta Hort 510: 37-44.

Kho YO, Den Nijs, APM, Franken J (1980) Interspecific hybridization in *Cucumis* L. 2. The crossability of species, an investigation of *in vivo* pollen tube growth and seed set. Euphytica 29: 661-672.

Koo D, Hur Y, Jin D, Bang J (2002) Karyotype analysis of a Korean cucumber (*Cucumis sativus* L. cv. 'Winter Long') using C-banding and bicolor fluorescence *in situ* hybridization. Mol Cells 13: 413-418.

Lv J *et. al.* (2012) Genetic diversity and population structure of cucumber (*Cucumis sativus* L.). PLoS ONE 7(10): 46919.

Perl-Treves R, Zamir D, Navot N, Galun E (1985) Phylogeny of *Cucumis* based on isozyme variability and its comparison with plastome phylogeny. Theor. Appl. Genet. 71: 430-436.

Ramachandran C, Seshadri VS (1986) Cytological analysis of the genome of cucumber (*Cucumis sativus* L.) and muskmelon (*C. melo* L). Z Pflanzenzuecht, 96: 25-38.

Sebastian P, Schaefer H, Telford IRH, Renner SS (2010) Cucumber (*Cucumis sativus* L.) and melon (*C. melo*) have numerous wild relatives in Asia and Australia, and the sister species of melon is from Australia. Proc Natl Acad Sci U S A 107, 14269–14273.

Staub JE, Serquen FC, McCreight JD (1997) Genetic diversity in cucumber (*Cucumis sativus* L.): III. An evaluation of Indian germplasm. Genet Resour Crop Evol 44: 315–326.

Staub JE, Fredrich L, Marty TL. Electrophoretic variation in cross compatible wild diploid species of *Cucumis*. Can J Bot 1987; 65: 792-798.

Staub JE, Knerr LD, Holder DJ. (1992) Phylogenetic relationships among several African *Cucumis* species. Can J Bot 70: 509-517.

Tagashira N, Pl¹der W, Filipecki M, Wisniewska A, Gaj P, Szwacka M, Fiehn O, Hoshi Y, Kondo K, Malepszy S (2006) The metabolic proiles of transgenic cucumber line vary with different chromosomal locations of the transgene. Cell Mol Biol Lett 10: 697-710.

Wehner TC, Guner N (2004) Growth stage, flowering pattern, yield and harvest date prediction of four types of cucumber tested at 10 planting dates. Proc. XXVI IHC. Advances in Vegetable Breeding (Eds) J.D McCreight and E. J Ryder Acta Hort 637.

Whitakar TW (1933) Bot. Gaz., 94: 780-90.

Zhuang FY, Chen JF, Staub JE, Qian CT (2004) Assessment of genetic relationships among *Cucumis* spp. by SSR and RAPD marker analysis. Plant Breeding 123: 167-172.

Chapter 2

Breeding for Downy Mildew Resistance in Cucurbits

N. Shilpashree, M. Anjanappa, D.C. Manjunathagowda, M. Prakash and J.C. Bommesh

ABSTRACT

Cucurbits are important vegetable crops of the family cucurbitaceae, these crops grown extensively in India. However, the productivity of cucurbits reported as low as 8-10 t/ha, it could be due to incidence of pests and diseases. Among them, downy mildew is an economically important disease which causes yield losses up to 79.06 per cent. The chemical method of management of this disease is costly, besides hazardous residual effects on the consumers. Therefore, it is imperative to concentrate on the development of varieties/hybrids with high level of resistance to downy mildew disease.

Keywords: *Cucurbits, Downy mildew, Resistance, Diseases, Varieties.*

Introduction

Cucurbits are important vegetable crops belong to a family *cucurbitaceae*, extensively grown in India, tropical and subtropical regions of the world. These crops cultivated by several parts of India (Uttar Pradesh, Punjab, Rajasthan, Karnataka and Andhra Pradesh). In Karnataka mainly cultivated cucurbits are cucumber, gherkins and melons. The fruits of cucurbits used as desserts (muskmelon and watermelon), cooking (gourds and squashes) and salads (cucumber and gherkins). These cucurbits are rich source of minerals and vitamins, so plays an important role in many health benefits of human body mainly for constipation and obesity.

Cucurbits are important vegetable crops grown extensively in India. However, productivity of cucurbits is low and it's attributed to prevailing pest and diseases. Among them, downy mildew is important disease which causes yield loss up to 79.06

per cent (Mahishri and Siradhara, 1988). The chemical method of management of this disease is costly besides hazardous residual effects on the consumers. Therefore, it is imperative to concentrate on the development of varieties with high level of resistance to downy mildew.

Downy mildew of cucurbits caused by an oomycetes fungi *Pseudoperonospora cubensis* (Berk. and Curtis.) Rostow. It has been reported in more than 70 countries worldwide (Lebeda and Cohen, 2011). It is most devastating disease and significantly affects the production of cucurbits. This species has distinct host specificity attacking above ground parts, mainly leaves. *Pseudoperonospora cubensis* was first described by Berkeley in 1868 in herbarium plant material originated from Cuba and hence its species name *cubensis*. It has a wide host range, infecting approximately 20 different genera of cucurbits. In India downy mildew on cucumber was first reported from Coimbatore (Tamil Nadu).

Symptoms of Downy Mildew

Downy mildew is a foliar disease and it exhibits symptoms mainly on leaves. Initially the infected leaves develop green to yellowish and dark green areas reflecting a mosaic pattern. These areas then transformed into yellow coloured angular spots restricted by the veins, which later increase in number and size. In

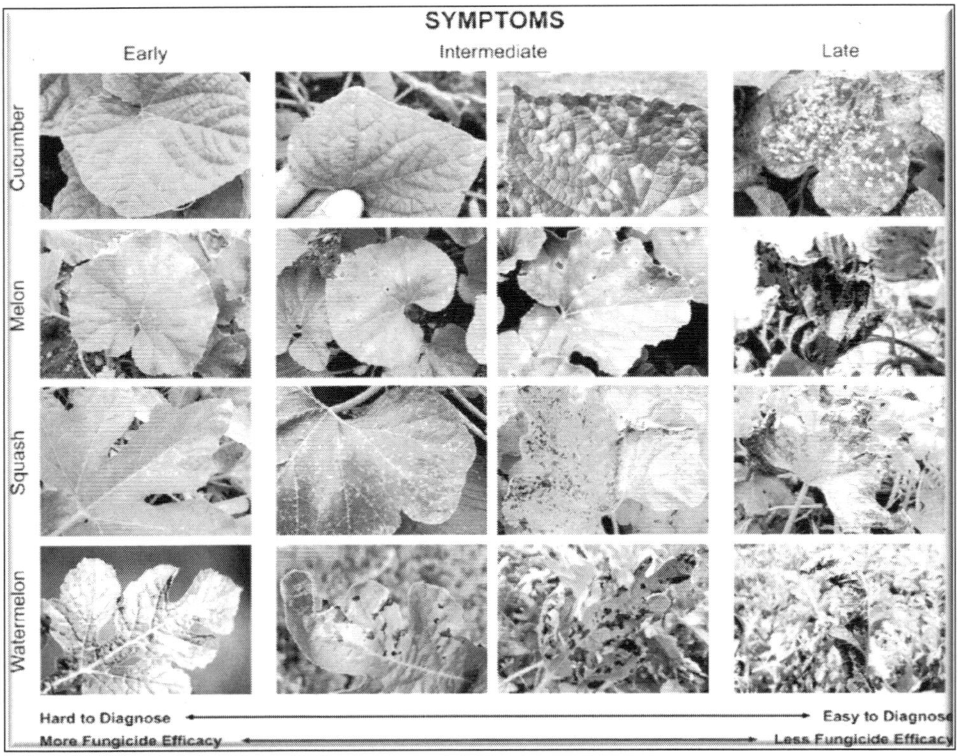

Figure 2.1. Symptoms of Downy Mildew in different Cucurbitaceous Crops.

the early stages, corresponding abaxial surface shows discrete cobweb like fungus growth visible only during early morning hours and cloudy weather. Ultimately, the severely infected leaves become chlorotic, turn brown and shrivel. The whole vine may wilt under severe disease intensity. As a result of early infection, cotyledons are also invaded. Five to fifteen days old leaves are more prone to infection.

Etiology

The downy mildew pathogen *P. cubensis* is a weather sensitive pathogen. Sporangia are the main source of inoculum produced through sporangiophores which are dichotomously branched at acute angle and these are easily dislodged by air and rain splash. Environmental factors have great impact on symptom expression, disease progress and epidemiology. In case of Moist humid weather conditions, the diseases infection starts and spreads very rapidly. Among the weather factors, temperature and leaf wetness greatly influences the degree if infection and initiation of disease. The optimum temperature for germination of sporangia and initiation of infection is 20-25°C during the day and 10-15°C during the night with minimum 2 h. wetting period of leaf. The incubation period depends on temperature, photoperiod, inoculum concentration and leaf wetness duration, and can range from 4 to 12 days (Lebeda and Cohen, 2011). Symptomatic plants with yellow lesions have the greatest sporulating capacity. The sporulating capacity of purely necrotic lesions is low and that of yellow-necrotic lesions is intermediate. Warm and dry weather, typical mist in early morning hours, causes twisting of the sporangiophores, which may be of importance of the detachment of sporangia.

Factors Associated with Downy Mildew Resistance

Effect of Pathogen

Resistance is associated with inoculum density and decreases in concentration of inoculums concentration decreases downy mildew incidence.

Temperature

Optimum sporangia production occurs between 15°C-20°C and requires atleast 6 hours of high humidity. Disease may progress slowly or stop temperature rise above 30°C during the day time.

Moisture

Honey dews and fog increases the disease incidence.

Rainfall

Heavy rain fall or precipitation increases the disease incidence.

Screening Technique for Downy Mildew Disease Resistance

Screening can be done by both natural and artificial inoculation methods. Artificial inoculation includes, detached leaf technique and leaf disc inoculation.

Artificial Inoculation

The downy mildew fungus *P. cubensis* was maintained on susceptible cultivar in growth chamber at 20°C. Freshly produced sporangia were harvested into cold (4°C) double distilled water and their concentration was adjusted to 1000 sporangia/ml with aid of a cytometer. Sporangia suspensions were used for foliar spraying of intact plants. Sprayed plants were kept in a dew chamber at 18°C in the dark for 20 hrs and incubated at 20°C (12 hrs light/day, 100 $\mu Em^{-2} S^{-1}$)

Artificial Inoculation using Leaf Disc Method

Leaf disc (22mm in diameter) were placed on 0.5 per cent water agar in petri dishes. Abaxial surface up and each was inoculated four 10 µl droplets of inoculums suspension containing approximately 70 sporangia per droplet. Dishes were incubated for 20 hrs at 18°C in the dark and then at 20°C.

Inoculation in the Field

Seedlings were raised in potting trays containing sterilized coco-peat as a growth media in greenhouse. As two leaf stage, seedlings were transplanted to main field. Seven days after transplanting, the adaxial leaf surface was sprayed with sporangial suspension containing 10,000 sporangia/ml by hand sprayer. Susceptible variety was planted at regular intervals all over the field for uniform spread of disease. (Shashikumar *et al.*, 2010)

Mechanism of Resistance

1. **Morphological feature:** Various morphological characters such as few, small sunken and hairy stomata with lesser opening duration and lesser opening duration of flower have been found associated with disease resistance in different crop plants.
2. **Hypersensitivity:** Hypersensitivity of the cells leading to cell death or chlorosis around the point of infection and consequent starving of the obligate parasite. This delays start of an epidemic and contributes to a defense mechanism against specific race of a pathogen.
3. **Biochemical features:** Toxic and potentially therapeutic compounds in cucurbits include, oxygenated tetracyclic triterpiniods such as cucurbitacin, saponin, glycosides, alkaloids, amino acids, phenols and xanthophyll's.

Cucumber

"The identification of sources of resistance to downy mildew disease in cucumber (*Cucumis sativus* L.) was to identify the downy mildew resistance lines in cucumber (Pitchaimuthu *et al.*, 2012). The experimental materials consisted of 42 accessions of *Cucumis spp* including the wild type *Cucumis hardiwickii, Cucumis sativus* var *sativus* and SM 12735 (Bitter in taste). This study also included a commercial cultivar Swarna Agathi which was used as susceptible check. Screened the lines using a field adapted method, diseases inoculation methodologies and scoring Percent Diseases Index (PDI) of a disease assessment scale for evaluating the resistance downy mildew in cucumber developed by Swamy *et al.* (1980) and Pitchaimuthu *et al.* (2007)". The PDI was calculated using the formula.

$$PDI = \frac{\text{Sum of numerical values}}{\text{Number of leaves graded x Maximum ratings}} \times 100$$

"After calculating the PDI, the germplasm in the population was categorized into four groups namely resistant (0-25), moderately resistant (26-40), susceptible (41-60) and highly susceptible (>60). In open field screening, wild species *Cucumis hardiwickii*-14 and 15, *Cucumis sativus* var *sativus* and SM 12735 exhibited a high level of resistance to downy mildew diseases with 0-25 per cent PDI undernatural conditions. Five accessions namely (IIHR-27, IIHR-35, IIHR-303, IIHR-64 and 82) had < 40 per cent PDI exhibited moderately resistance to PM and DM diseases and rest of the accessions showed > 60 per cent PDI depicting highly susceptible to both the diseases. This preliminary work must be repeated by screening all lines under artificial conditions in the glasshouse in order to confirm the results".

The objective of this study is to evaluate the germplasm collected from National Bureau of Plant Genetic Resources (NBPGR) New Delhi, India to identify new resistant genotypes for downy mildew pathogen of cucumber (Bommesh *et al.*, 2018). The experimental materials consisted of 41 genotypes obtained from the National Bureau of Plant Genetic Resources, New Delhi, which includes indigenous collections (36), exotic collections (3) and one wild *Cucumis* species. They were evaluated for resistance to downy mildew and Swarna Agethi was maintained as susceptible check. The disease scoring was started a month after planting at weekly interval for 7 weeks until the end of the crop growth. Disease incidence was calculate by 0-9 scale then based on this data Per cent Disease Incidence (PDI) was calculated, the genotypes were categorized into four groups namely resistant (0–20 per cent), moderately resistant (21–40 per cent), susceptible (41–60 per cent) and highly susceptible (> 60 per cent).

Artificial Screening using Detached Leaf Assay

Leaves were collected from plants heavily infected leaves were soaked in distilled water and rubbed gently with a glass rod to dislodge the sporangia. The abaxial surface of leaf was kept up and each leaf inoculated with 8–10 droplets of the solution (6 ml each). The Petri dishes with the inoculated leaves were immediately placed in a growth chamber at dark at 20°C for 12 h and then kept in light for 10 days. Disease progress was assessed after the inoculation. Disease incidence data recorded at weekly interval for 7 weeks was used to calculate the apparent infection rate and then the sporangia were enumerated using haemocytometer. Disease incidence data recorded at a weekly interval was used to calculate the area under disease curve progression (AUDPC).

Among 41 genotypes, none of the genotype was observed to be immune to downy mildew pathogen. Genotype IIHR-438 and wild cucumber (*C. metuliferus* L.) showed resistance with average PDI of 17.66 and 17.46, respectively the response of resistant genotypes in cucumber is characterized by the recognition (compatibility/

incompatibility) of a host plant by its oomycetes pathogen. The resistant genotype IIHR-438 and wild cucumber (*C. metuliferus* L.) had 3.1 and 3.3 average number of spores per cm^2, respectively whereas Swarna Agethi (Susceptible check) had 3.80 spores (Figure 2.2). There was increase in the infection rate of downy mildew growth observed per day in all the genotypes and thus the chance of causing epidemics was more in almost all the genotypes of cucumber. The highest average 'r' value was observed in the susceptible genotypes of IIHR-389 and IIHR-393 is 1.23. *C. metuliferus* (0.74) and IIHR-438 (0.79) showed the least 'r' value as compared to susceptible check Swarna Agethi (1.20).

Figure 2.2. Downy Mildew Infected Plants in Susceptible Checks and Resistant Genotype Screened under Natural Conditions.

Ten genotypes showing resistance and high susceptibility were selected for the detached leaf assay by artificial inoculation method to confirm the resistance. Average PDI of *C. metuliferus* and IIHR-438 was 12.8 and 14.3, respectively. Resistant genotypes found resistance reaction under both field and artificial condition (Figure 2.3). Both the resistant genotypes IIHR-438 and *C. metuliferus* L. showed less AUDPC value of 772.24 and 764.49 respectively, compared with other genotypes under screened for resistance.

The results of logistic model for downy mildew resistance are explained. Parameter estimate of fitted models, measures of goodness of fit of logistic model R^2 and MSE) along with tested measures of model adequacy and AUDPC are also represented. Result indicates that the severity of downy mildew infection during plant growth can be explained with a logistic model. The accuracy of 99.9 per cent and 94.60 per cent in cucumber genotype IIHR-417 and IIHR-438 with error mean square of 1.08 and 8.19 respectively. In run test stat value, assumptions about residuals showed that error is distributed well within the critical region for all 41 genotypes. In the non-linear model the intrinsic growth rate 'a' was in the range of 0.17 (IIHR-177) to 1.40 (IIHR-438) and carrying capacity 'C' (Maximum disease growth rate) was minimum in resistant genotype IIHR-438 whereas susceptible check had comparatively high C value of 188.76. The 't' value in resistant genotype was 19.44 per cent in fourth week, whereas in susceptible variety it was 57.64 per cent in the third week itself. From this study, it was confirmed that the genotypes IIHR-438 and wild cucumber (*C. metuliferus* L.) showed resistance with less disease progression and can be utilized in breeding programs for disease resistance.

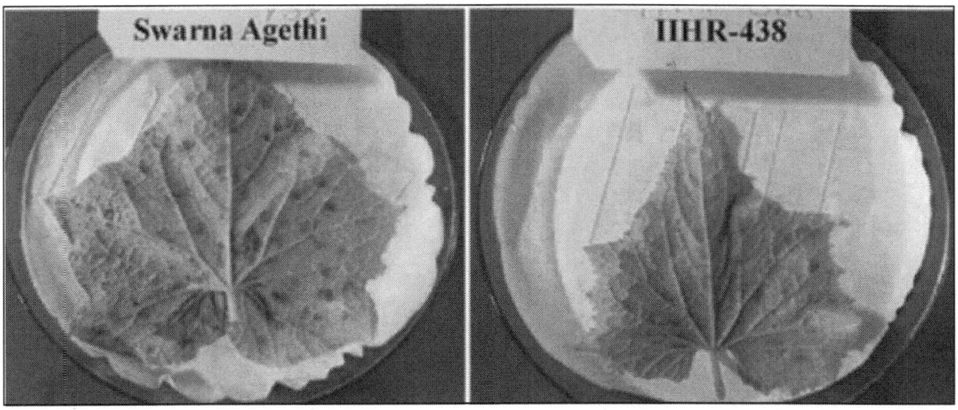

Figure 2.3. Infection of Downy Mildew in Susceptible Checks and Resistant Genotype by Detached Leaf Assay.

In case of gherkin cucumber, the study determined the pathotype of downy mildew isolates collected at ATRI, Lampang province (Charoenwattana *et al.*, 2017). The screening technique was used to screen and select gherkin cucumber lines under greenhouse and field conditions to increase plant resistance to the pathogen using the pedigree method. Isolates of *P. cubensis* were collected from downy mildew lesions on heavily infected cucumber leaves at ATRI, Lampang province (Lampang isolate) to determine host specificities of the pathogens. Inoculations were made to 15 cultivars representing 8 species within 5 genera of the family *cucurbitaceae*, with 12 plants per cultivar per assay and repeated twice. Once they reached a two-expanded-cotyledon stage, adaxial and abaxial leaf surfaces were inoculated with a suspension of *P. cubensis* (104 sporangia ml^{-1}) by spraying until incipient run-off. Inoculated plants were placed in darkness at 20°C with high humidity for 21 hours. Subsequently, they were transferred to a greenhouse with a temperature of 24-28°C and a relative humidity of 80-90 per cent. The plants were observed daily for symptom development. Inoculated plants were evaluated for the presence of lesions and sporulation intensity. Sporulation was rated qualitatively as compatible or incompatible. The disease reactions of the host plants were recorded on day seven after inoculation. The host plants without fungal inoculation were used as a control. To improve resistance to *P. cubensis* in gherkin cucumber lines, four generations of gherkin cucumber lines (F_2-F_5) were tested and screened, along with gherkin varieties, commercial varieties of cucumber and downy mildew resistant varieties using a screening technique. The pathogenicity test in this study originated from 169 gherkin cucumber lines of the F_2 generation. The disease resistance of the F_3 and F_4 generations were tested and screened in the greenhouse. The F_2-F_5 generations of gherkin cucumber lines were also tested and screened under field conditions. All cucumber lines and varieties were rated for resistance to downy mildew after inoculation for foliar lesions with a 0-5 visual rating scale (0–no foliar symptoms, 1–1-20 per cent symptoms, 2–21-40 per cent symptoms, 3–41-60 per cent symptoms, 4–61-80 per cent symptoms, 5–81-100 per cent symptoms) with some modifications for greenhouse and field assessment.

Inoculations of *P. cubensis* (Lampang isolate) to 15 cultivars representing 8 species within 5 genera of the *cucurbitaceae* family showed physiological reactions of the Lampang isolate and was classified as pathotype 3. This isolate was highly compatible with *Cucumis sativus, C. melo* var. *reticulatus, C. melo* var. *conomon* and *C. melo* var. *acidulous*. A pathogenicity test showed that a host specificity of *P. cubensis* isolate could be used to identify downy mildew in cucumber.

The screening technique applied to the F_5 generation using pedigree selection in the field produced 8 elite lines (out of 52) of gherkin cucumbers with an average score of 1.7, compared to an average score of 2.7 for gherkin varieties, 1.9 for commercial varieties of cucumber and 1.7 for downy mildew resistant varieties after 40 days of transplanting. These gherkin cucumber lines (P1-P8) can be used as parental lines to generate a hybrid variety and perform a combining ability test. These 8 elite lines from the F_5 generation yielded an average output of 26.9 tons per hectare, produced 32.6 fruits per plant and weighed 19.4 grams per fruit (compared to 23.1 tons per hectare, 27.5 fruits per plant and 19.8 grams per fruit for gherkin varieties). The cucumber downy mildew (*P. cubensis* - Lampang isolate) was identified as pathotype 3 for the presence of highly compatible reactions on four specific cucurbit host species. The results indicated that the screening technique used in the pedigree method increased downy mildew resistance and thus outputs of gherkin cucumber lines.

Cucumber accessions were screened for downy mildew resistance under field condition for four crop cycles at Rajamangkala University of Technology Lanna, Lampang, Thailand during 2006-2009 (Innark *et al.*, 2014). Disease score was rated 45 days after planting and each accession was classified into five classes based on the average of percentage of symptomatic leaf area. Cucumber accession was classified as highly resistant when the symptom was less than 20 per cent; moderately resistant whenthe symptom was between 20 and 40 per cent; intermediate when the symptom was between 40 and 60 per cent; moderately susceptible when the symptom was between 60 and 80 per cent and highly susceptible when the symptom was more than 80 per cent. Seeds were germinated on culture tray filled with soil and grown at 25°C until the cotyledons expand. Then, the cotyledons were used for DNA extraction using the modified CTAB protocol. The DNA samples were diluted (100 ng/ml) for ISSR analysis and stored at -20° C until used. A set of 20 ISSR primers were used for screening in this study for two replications. In order to evaluate the downy mildew disease reaction of forty cucumber accessions, natural infection in the field was used for four crop seasons from 2006 to 2009. The results showed that there was one highly resistant accession (PI489752), 23 moderately resistant accessions, three intermediate accessions, seven moderately susceptible accessions, and six highly susceptible accessions. The average of per cent of symptomatic leaf area ranged from 20.1 per cent (PI489752) to 95.0 per cent (PI606014) and the standard derivation (SD) ranged from 1.7 (PI432878) to 44.5 (PI212985). To investigate the genetic diversity of downy mildew resistant/susceptible cucumber germplasms, 20 ISSR primers were used to screen 40 cucumber accessions. Among 20 ISSR primers, 17 ISSR primers (85 per cent) could amplify all 40 cucumber accessions while three ISSR primers (15 per cent) could not amplify. Eighty four ISSR bands were produced, of which 16 bands (19 per cent) were monomorphic and 68 bands (81 per cent) were polymorphic. The utility of ISSR primer was measured by polymorphism information content (PIC) and resolving power (RP). The PIC values were calculated by the mean of PIC valued for all fragments of each ISSR primer. As result, the PIC value ranged from 0.12 (UBC-811) to 0.45 (UBC-823). The mean of PIC value was 0.25. The RP is the ability to distinguish genotypes. In this study, UBC-849 had thelowest RP value (0.5) while UBC-824 had the highest RP value (3.20). The mean RP of 17 ISSR primers were 1.82.

According to the dendrogram, these three clusters could be classified by both downy mildew disease reaction, and country of origin. In term of downy mildew disease reaction, cluster A and B exhibit downy mildew resistance (highly and moderately resistant) while cluster C exhibit downy mildew susceptible (highly and moderately susceptible). Moreover, the country of origin of cucumber could be divided from this tree, which cluster A originated from East Asia, cluster B originated from Thailand, and cluster C originated from India (Figure 2.4).

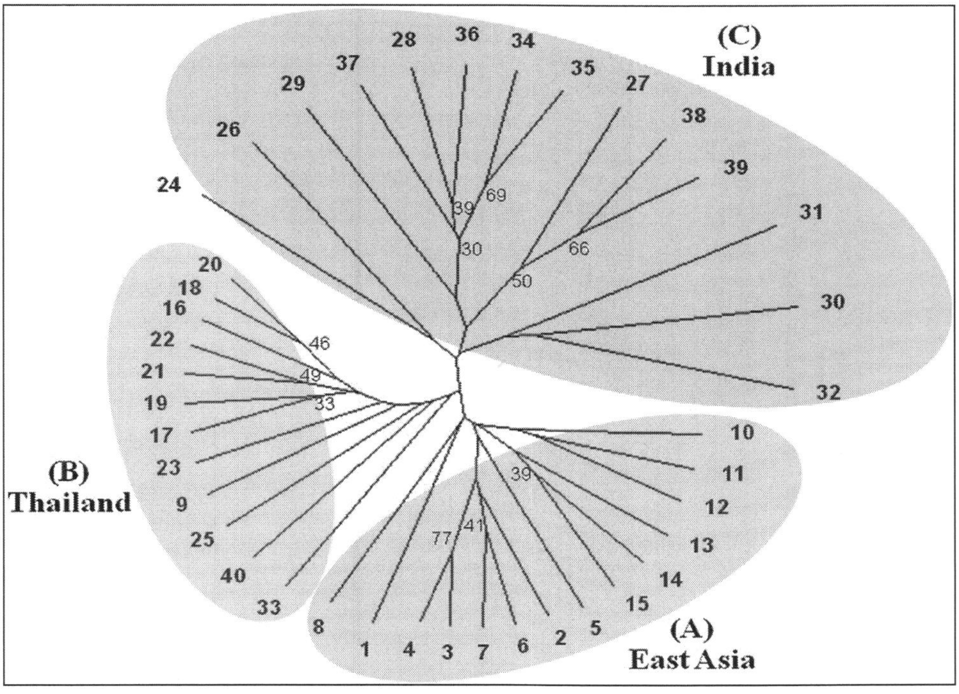

Figure 2.4. The Phylogenetic Tree of 40 Downy Mildew Resistant/Susceptible Cucumber Germplasm was Constructed using 20 ISSR Markers by UPGMA Method.

The genetic diversity of 40 downy mildew resistant and susceptible germplasm was evaluated using 20 ISSR markers. The constructed dendrogram could be divided by downy mildew disease reaction and country of origin. Moreover, yield and yield components of cucumber could also be divided by this dendrogram which are the beneficial to cucumber breeders in the process of parental selection for cucumber breeding program to improve the resistant to downy mildew and high yield cucumber varieties in the future.

The study, QTL mapping of downy mildew resistance in an introgression line derived from interspecific hybridization between cucumber and *Cucumis hystrix* conducted to identify regions of the genome with QTLs for resistance to downy mildew and identified molecular markers linked to QTLs (Pang *et al.*, 2013). Two cucumber cultivars and 10 ILs derived from backcrossing the *Cucumis hystivus* to cucumber were used. An F_2 population was developed from a cross between the downy mildew resistant IL52 and downy mildew susceptible cucumber 'Changchunmici'. 300 F_2 individuals were obtained by selfing the selfing the F_2 progeny. Cultivars NCSUM21 and WisSM18 were used as resistant and susceptible control respectively. Following three phenotypic identification methods: yellowing, leaf necrosis and general health traits, IL52 was regarded as highly resistant and used as a resistant parent to make the F_2 population with 'Changchunmici'. Construction of the genetic linkage map from the set of 300 F_2 plants developed from a cross between 'Changchunmici' and IL 52, comprising 100 SNP markers and four SSR markers. The map comprised seven chromosomes and spanned a genetic distance of 6.2 cM among markers (Figure 2.5). The genetic map was used to perform QTL analysis for yellowing, necrosis and general health traits for downy mildew. Interval mapping analysis for DM-resistance identified two QTLs on chromosome 5 (DM_5.1 and DM_5.2). DM_5.1 and DM_5.2 were supported by LOD scores of 9.66 and 7.48 and explained 17.9 and 14.2 per cent of the variation, respectively. Two QTL for DM-yellowing were located in the same regions as DM-resistance, showing LOD scores (15.04, 7.25) and explaining 26.8 per cent and 14 per cent of the variation, respectively. For necrosis, one QTL was determined on chromosome 5 (Necr_5.1) that explained 18.3 per cent of the variation and one on chromosome 6 (Necr_6.1) that explained 13.9 per cent of the variation (Figure 2.5).

This study inferred that the report ofQTL mapping of downy mildew resistance in a cucumber introgression line derived from a cross with *Cucumis hystrix*. Three QTL regions were found in the introgression line IL 52, which could be used for pyramiding of downy mildew resistance to provide durable protection.

Muskmelon

Shashikumar *et al.* (2010) conducted a study to determine the gene action which governing the inheritance of downy mildew resistance in muskmelon. Downy mildew resistant lines IIHR 121 and IIHR 122 were derived from segregating melon lines introduced from Russia, EC 564749 and EC 564750, respectively which both belongs to *Cucumis melo* var. *cantaloupensis*. The lines were purified for five generations before use in this study. IIHR 121 and IIHR 122 were crossed with downy mildew susceptible variety Punjab Sunehri (Punjab Agriculture University,

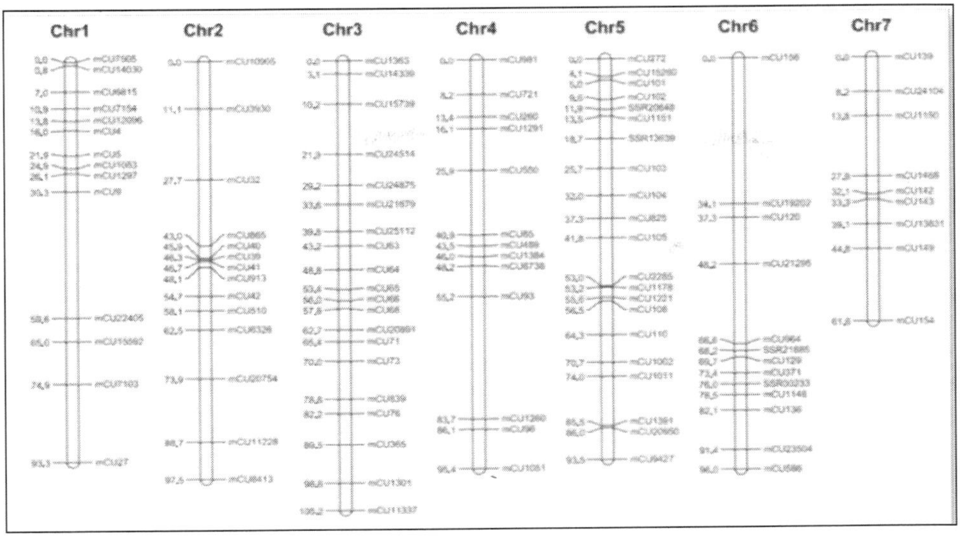

Figure 2.5. Genetic Linkage Map Based on the F_2 Population Derived from the Cross Changchunmici x IL52.

Ludhiana, India), cultivar RM 43 (popular cultivar in Northern states of India, highly resistant to powdery mildew) and inbred line IIHR 681 (belongs to species *Cucumis callosus*, resistant to fruit fly) to derive four F_1 hybrids *viz.* Punjab Sunehri 9 IIHR 122, RM 43 9 IIHR 121, IIHR 681 9 IIHR 121 and IIHR 681 9 IIHR 122. These F_1's were crossed to their parents P_1 (susceptible) and P_2 (resistant) to get BC_1P_1 and BC_2P_2 generations, respectively. On the same F_1's, F_2 seeds were generated by self-pollination. The experimental material comprised of six generations (P_1, P_2, F_1, F_2, BC_1P_1 and BC_2P_2) derived from each of the four crosses was screened in field and greenhouse conditions for resistance to downy mildew during 2006–07 at the Indian Institute of Horticultural Research (IIHR), Bangalore, India. The F_1 mean was significantly lesser than the midparent value in all the crosses. In both screening experiments and all crosses, the mean of F_2 exceeded that of the F_1 and BC_2P_2 but lower than BC_1P_1 generations. In general, mean of BC_2P_2 generations were lesser than the mean of F_1 generations in all the crosses. Individual F_2, BC_1P_1 and BC_2P_2 progeny observed that transgressed with extreme mean PDI values than either parent in both screening experiments. This could be due to contrasting parental means. However, contrasting parents are prerequisite for generation mean analysis.Scaling tests indicated presence of epistasis in all the crosses. All the six genetic parameters were significant in all crosses except RM 43 9 IIHR 121 in both screening experiments. All the crosses expressed significant and positive additive (*d*) gene effects in both experiments. In field experiment, dominance x dominance (*l*) interaction effects was the largest component of gene effects in three out of four crosses. Dominance (*h*) and dominance x dominance (*l*) gene effects signed opposite in all crosses and both experiments. However, in greenhouse experiment, significant and negative additive 9 additive effects expressed in crosses IIHR 681

9 IIHR 122 and IIHR 681 9 IIHR 121. The sum of additive effects (*d+i*), in terms of fixable component was much lower than the sum of dominance effects (*h+l*) in terms of non-fixable component in all the crosses in both experiments. However, negative sign associated with gene effects indicated, in those crosses, disease level could be decreased in relation to midparent. The estimates for broad-sense heritability ranged from 0.86 to 0.92 averaging 0.88 in field experiment and 0.74 to 0.88 averaging 0.81 in greenhouse experiment, respectively. Two different screening experiments revealed that fixable gene effects (*d*) and (*i*) were lower than the non-fixable (*h*) and (*l*) gene effects in all the crosses indicating greater role of non-additive effects in the inheritance of resistance to downy mildew. Breeding methods like diallele selective mating or biparental mating in early segregating generations might prove to be effective approaches. However, high mid-parent heterosis indicated that the resistance is controlled mainly by dominance effects, therefore the inbred lines IIHR 121 and IIHR 122 could be used strategically to exploit heterotic effects.

Ridge Gourd

Thammaiah *et al.* (1999) screened ridge gourd genotypes against downy mildew at different locations. In Bijapur conditions, Raichur Local-2 showed less downy mildew incidence (22.20 per cent) on leaf area affected followed Deodurge Local (26.93 per cent) leaf area affected. In Arabhavi conditions, Deodurge Local (3.4 per cent) followed by Govnal Local (4.2 per cent) whereas Raichur Local-2 recorded (9 per cent) was lowest incidence. Narasannavar *et al.* (2018) was carried out an experiment in Department of Vegetable Science, COH, Bengaluru. Total 41 genotypes were used including resistant check (Deepthi) and susceptible checks (Pusa Nutan and Pusa Nasdar). These were subjected to natural field conditions. Symptoms of downy mildew were recorded 45, 60, 75 and 90 days after sowing as per the scale (0-5 scale). In open field screening for downy mildew has led to identification of two resistant genotypes (COHB-9 and COHB-32), two moderately resistant lines (COHB-10 and COHB-40) and one was highly susceptible (COHB-8) whereas remaining genotypes were exhibited moderately susceptible and susceptible reaction to downy mildew in ridge gourd. Therefore these genotypes could be utilized as parent in hybridization programme to incorporate the downy mildew resistance under the background of high yielding and good market quality.

Bottle Gourd

"The different varieties of bottle gourd were screened to identify downy mildew resistant variety (Uraiha *et al.*, 2018). Thirteen varieties of bottle gourd were evaluated for resistance against downy mildew disease during *rabi* 2017. The severity of downy mildew was recorded for each variety at 30, 60 and 90 days after transplanting. The severity and location severity index (LSI) were calculated by using the formulae as given below. Thereafter, all the varieties were grouped in different reaction categories by using the 0-5 scale".

$$\text{PDI} = \frac{\text{Total leaf tissue damages}}{\text{Total healthy tissue of leaf}} \times 100$$

$$LSI = \frac{\text{Sum of multiplication of variety and scale}}{\text{Total number of variety}} \times 100$$

"Among the thirteen varieties of bottle gourd evaluated three varieties *viz.* Amrit, Ankit and Anmol were free from the disease and showed immune reaction against disease. However, two (Manya, NS-433), four (Chutki, Haruna, Latto and Naveen) and three (Angad, Mahima and Sarita) varieties were resistant, moderately resistant and moderately susceptible, respectively. One variety *i.e.* Divya showed susceptible reaction against downy mildew while none of the variety was found highly susceptible. Out of thirteen varieties of bottle gourd tested five (Amrit, Ankit, Latto, Manya and NS-443) and two (Manya, NS-433) varieties were found successively immune and resistant against downy mildew".

Bitter Gourd

Pandey *et al.* (2005) studied on reaction of *P. cubensis* after challenge inoculation on bitter gourd. Out of 142 genotypes/germplasm lines screened against downy mildew, VRBT-39, NIC-12285 were moderately resistant with high degree of tolerance. IC50524A, Wild-27, VRBT-1, IC68250B, IC683013, IC85618, VRBT-21 were found moderately susceptible.

Pumpkin and Snap Melon

Khanal (2017) conducted an experiment on Identification and characterization of resistance against downy mildew in pumpkin (*Cucurbita moschata* L.) and snap melon (*Cucumis melo* L. var. *momordica*). The role of peroxidases, as a group of pathogenesis-related (PR) proteins, which are involved in lignin production in phenylpropanoid pathway were investigated in the interaction of snapmelon and pumpkin accessions with *P. cubensis*. A possible function of peroxidases is the formation of structural barriers such as cell wall enhancement and deposition of cell wall appositions, both of which can be involved in the polymerization of lignin or suberin, the cross-linking of wall glycoproteins or polysaccharides, and the dimerization of antimicrobial phenols.

Activity of peroxidase was significantly higher in resistant accessions of snapmelon and pumpkin as compared to susceptible and highly susceptible accessions. Activity was higher after 7 days after inoculations followed by 14 and 21 days after challenge inoculation. Mean peroxidase activity was the highest in resistant accession of snapmelon MC 2013-4 and pumpkin accession Punjab Samrat *i.e.* (18.96 ΔA/min/g FW) and (17.21 ΔA/min/g FW) respectively. However, activity declined in highly susceptible accessions SM 2013-13 (snapmelon) and BN-364 (pumpkin) *i.e.* 6.33 and 10.33 ΔA/min/g FW respectively. Thus peroxidase activity remained higher till 7 days after the challenge inoculation of *P. cubensis*.

Conclusion

The development of downy mildew disease resistant varieties and hybrids with high yield and quality is an immense scope in near future. As the developed varieties and hybrids with resistance to downy mildew disease may become susceptible in due course of time due to evolution of new races of pathogen in nature. Hence, the continuous efforts should be made to develop new varieties and hybrids with horizontal resistance which will be economically feasible.

REFERENCES

Bommesh JC, Pitchaimuthu M, Sadashiva AT, Sriram S, Varalakshmi B, Ravishankar KV. Identification and confirmation of downy mildew (*Pseudoperonospora cubensis* Berk. and Curt.) resistance sources in cucumber (*Cucumis sativus* L.). Indian Phytopathol., 2018; 71(3), 337-348.

Charoenwattana P, Khanobdee C, Udomyotin A. Screening techniques for downy mildew resistance in gherkin cucumbers. Int. J. Geomate., 2017, 13(40), 35-42.

Dhiman JS, Lal T, Bajaj KL. Evaluation of muskmelon genotypes for multiple disease resistance, yield and quality parameters. Trop. Agric., 1995; 72(1): 58-61

Innark P, Ratanachan T, Khanobdee C, Samipak S, Jantasuriyarat C. Downy mildew resistant/susceptible cucumber germplasm (*Cucumis sativus* L.) genetic diversity assessment using ISSR markers. Crop Protection, 2014; 60,56-61.

Khanal S. Identification and characterization of resistance against downy mildew in pumpkin (*Cucurbita moschata* L.) and snapmelon (*Cucumis melo* L. var. *momordica*), Dissertation: M. Sc. (Agri.) Thesis, Punjab Agric. Univ. Ludhiana, 2017.

Lebeda A, Cohen Y. Cucurbit downy mildew (*Pseudoperonospora cubensis*) - biology, ecology, epidemiology, host-pathogen interaction and control. European J. Plant Pathol. 2011; 129: 157-192.

Mahishri RE, Shiradara BS. Studies on downy mildew of cucurbits in Rajasthan: incidence, distribution, host range and yield loss in musk melon. Ann. Arid Zone. 1988; 27(1): 67-70.

Narasannavar A, Devappa V, Pitchaimuthu M, Fakrudin B, Lingaiah HB, Anjanappa M, Sriram S. Selection parameters for downy mildew disease, fruit yield and related traits in ridge gourd [*Luffa acutangula* (Roxb.) L.]. Electron. J. Plant Breed. 2017; 8(2): 601-608.

Pandey KK, Ram D, Pandey PK, Rai M. Artificial screening technique and sources of resistance to downy mildew disease in bitter gourd. V*e.g.* Sci., 2005; 32(1): 107-108

Pang X, Zhou X, Wan H, Chen J. QTL mapping of downy mildew resistance in an introgression line derived from interspecific hybridization between cucumber and *Cucumis hystrix*. J. Phytopathol. 2013; 161: 536-543.

Pitchaimuthu M, Swamy KRM, Girija Ganeshan. Evaluation of muskmelon hybrids for resistant to downy mildew (*Pseudopernospora cubensis*) disease paper presented on III National Symposium on plant protection in Horticulture held at IIHR Bangalore on 7-9 March 2007; p28.

Pitchaimuthu M, Souravi K, Ganeshan G, Kumar GS, Pushpalatha R. Identification of sources of resistance to powdry and downy mildew diseases in cucumber

[*Cucumis sativus* (L.)]. Pest Management in Horticultral Ecosystems. 2012; 18(1): 105-107.

Shashikumar KT, Pitchaimuthu M, Rawal RD. Generation mean analysis of resistance to downy mildew in adult muskmelon plants. Euphytica. 2010; 173: 121-127.

Swamy KRM, Dutta OP, Ullas BA, Identifying the sources of resistance to downy mildew and powdry mildew diseases in muskmelon. Proc. Nat. Sem. Dis. Res. Crop Plants, 1980, pp. 101-104.

Thammaiah N, Reddy BS, Patil RV, Kulakarni MS, Haralapur SI. Varietal screening of ridge gourd genotypes against downy mildew. South Indian Hort., 1999; 47(1-6): 315-316.

Uraiha V, Khare CP, Koshale KN, Tamin. Evaluation of different varieties of bottle gourd against downy mildew. Int. J. Chemical Studies. 2018; 6(6): 85-87.

Chapter 3

Breeding Strategies for YVMV Resistance in Okra

K.N. Kavya, D.C. Manjunathagowda, G. Prakash and R. Selvakumar

ABSTRACT

India is the leading producer of okra in the world, the production of okra has been decreased due to the severe incidence of Yellow Vein Mosaic Virus (YVMV) disease, transmitted by whitefly insect and cause yield loss (50-94 per cent). The wild taxa Abelmoschus crinitus, A. pungens, A. angulosus, A. manihot, A. caillei, and A.tetraphyllus were found resistant to YVMV, cross between YVMV resistant and susceptible cultivars resulted in transfer of genes and were brought together in the F_1 *hybrids, due to presence of single dominant gene controlling YVMV resistantance along with some minor genes. In the recent past, the resistant varieties became susceptible due to various factors. Hence, main objective of okra is to develop stable YVMV resistant varieties through suitable breeding approaches is very essential. Breeding for YVMV resistance with stable high yielding genotypes is possible by adopting both conventional and non-conventional approaches. Understanding the genetics of resistance in donor parent helps in choosing suitable breeding method for development of YVMV resistant okra varieties. Using resistance sources from wild-related species serves as the best sources of resistance to YVMV which is need to be exploited to develop stable viral-resistant varieties in okra in order to reduce marketable yield loss.*

Keywords: *Abelmoschus spp., Okra, YVMV, Resistance.*

Introduction

India is the leading producer of okra in the world and ranks first with a production of 5784.0 thousand tons (72 per cent of the total world production) of okra from over 498.00 thousand hectare areas. It is also cultivated across the world namely Nigeria, Sudan, Pakistan, Ghana, Egypt, Benin, Saudi Arabia, Mexico and Cameroon. Andhra Pradesh is the leading okra producing state which has production of around 1184.2 thousand tons from an area of 78.90 thousand ha with a productivity of 15 tons/ha followed by West Bengal of 862.1 thousand tons from 74.00 thousand ha with 11.70 tons/ha productivity (NHB, 2017).

Okra is an important source of vitamins, calcium, potassium and other minerals which are often lacking in the diet in developing countries. The crop is prone to damage by various insects, fungi, nematodes and viruses, although there is wide variability in their degree of infestation. Some of the important insect pests are shoot and fruit borer, jassid, aphid, whiteflies and mites *etc.* The okra is also subjected to the attack of many diseases affecting leaves, flowers and fruits. Its cultivation in India is challenged due to severe incidence of Yellow Vein Mosaic Virus (YVMV) and Enation Leaf Curl Virus (ELCV) diseases, spread by an insect vector namely white fly (*Bemisia tabaci*). The loss in marketable yield has been estimated at 50-94 per cent depending upon the stage of crop growth at which the infection occurs (Venkataramanappa *et al.*, 2011. 2012). Lack of sources of resistance to these viruses in cultivated species has forced breeders to look into the wild species for resistance sources. The transfer of resistance from wild relatives has also been hampered by sterility problems. Hence continuous search for new sources of resistance and development of better varieties/hybrids with higher level of resistance should be prime objective. In India, a number of ICAR Institutes, State Agricultural Universities and private seed companies are working on various aspects of genetic improvement of okra in order to develop high yielding and disease resistant varieties. Through intensive research efforts more than 50 improved varieties and hybrids have been released. Some of these varieties and hybrids have already made significant impact/ contribution in revolutionizing the production of okra in the country. However, still a vast gap exists in the research efforts and the expected outcome. Emerging biotypes of whitefly, new virus strains and breakdown of resistance in okra varieties/hybrids are the major challenges which need to be addressed.

Plant Viruses

The begomoviruses are most prevalent (25 per cent) plant virus followed by *potyvirus* (21 per cent), *tospovirus* (12 per cent), *crinivirus* (9 per cent) and *potexvirus* (8 per cent). In Asia 21 plant viruses are reported and among these 19 viruses are reported in India. Twenty-seven begomoviruses and one each of *ilarvirus*, *nepovirus*, *potyvirus*, *tospovirus* and *tymovirus* cause infection in okra and cause several viral diseases (Venkataramanappa *et al.*, 2011, 2012). In tobacco streak virus infected plant, fruits become distorted and this disease is transmitted by thrips and seeds. Yellow vein mosaic is most devastating disease of okra cause more than 60 per cent yield loss. Okra Enation Leaf Curl Virus is serious in North. TSV an ilarvirus is an emerging problem for okra production. Begomoviruses have high recombination rate and the presence of B-biotype whiteflies are contributing to epidemics of begomoviruses in okra. Host genetic resistance to viruses is one of the most practical, economical and environmentally secures strategies for reducing yield losses in okra. Begomoviruses possess a genome comprised of one or two circular ss-DNA molecules, genome is monopartite (2.5–3.0 kb) or bipartite (2.5–5.0 kb), having smallest known genome for an independently replicating twinned icosahedral virions [18x30 nm] and transmitted primarily by whitefly *Bemisia tabaci* and few are also transmitted mechanically. Natural resistance is available in wild species and land races of bhendi which has to be identified for each virus and ecological region. The symptoms of different viral diseases in okra, distribution of wild species of okra along with the begomoviruses in Indian sub-continent. In India okra crop is infected by seventeen types of begomoviruses. The PCR based diagnostic techniques of OYVMV, OELCV and TSV; Detection of BYVMV in whiteflies using PCR and techniques of Identification of vector biotypes (Venkataramanappa *et al.*, 2011. 2012). Resistance to yellow vein mosaic in okra is governed by recessive alleles, two complementary dominant genes and a single dominant gene in different resistant sources. On this account he also presented the different techniques of resistance breeding in okra. Efficient screening of genetic materials may be done through screening in hot spots/regions, artificial screening through whitefly and screening through grafting/agro-inoculation. Resistance evaluation can be performed through sensitive assay like visual symptoms, ELISA/nucleic acid probe and PCR assay. Punjab, Haryana, Andhra Pradesh, Jharkhand, West Bengal and Odisha were identified as hot spots for yellow mosaic disease screening. The information on artificial screening techniques for YVMV resistance and virus disease effectively utilized by indexing scale. Zero per cent disease incidence along with negative probe/PCR gives immune reaction, while 1-10 per cent and 11-25 per cent disease incidence with positive/negative and positives probe/PCR provide highly resistant and resistant, respectively. In concluding remarks he outlined the future plans *viz.*, identification of stable sources of resistance, identification of virus specific resistance, introgression of resistance genes from wild species, virus specific and source specific genetics of inheritance and identification of resistance to whitefly (Venkataramanappa *et al.*, 2011, 2012).

Multipartite Plant Viruses

Begomovirusses can either have a monopartite or bipartite genomes. Bi-partite begomovirus genomes consists of two ssDNA molecules of approximately 2.6 kb, referred to as DNA-A and DNA-B, with each components are involved in encapsidation and replication, whereas the genes on the B components are involved in viral systemic movement, host range determination, and symptom expression. Recently, some satellite DNA molecules, known as DNA-β or beta satellites, have

Table 3.1. Genome Segment is Packaged into a Separate Virus Particle (Venkataramanappa *et al.*, 2011. 2012)

Family	*Segments*
Begomovirus (Geminiviridae) (single-stranded DNA)	Bipartite/Monopartite
Comovirus (single-stranded RNA)	Bipartite
Furovirus (single-stranded RNA)	Bipartite
Tobravirus (single-stranded RNA)	Bipartite
Partitiviridae (double-stranded RNA)	Bipartite
Bromoviridae (single-stranded RNA)	Tripartite
Hordeivirus (single-stranded RNA)	Tripartite

been found associated with mono-partite begomoviruses. Like DNA-B, these satellites depends upon the DNA-A component for encapsidation, replication, and insect-transmission. Segmented virus genomes are those which are divided into two or more physically separate molecules of nucleic acid, all of which are then packaged into a single virus particle.

Table 3.2.Viruses Infecting Okra (Venkataramanappa *et al.*, 2011, 2012)

Virus group/Genus	*Transmission/ Vector*	*No. Virus Species*	
		World	*India*
A. *Begomovirus*	Whitefly	14	17
1. Bhendi yellow vein Bhuvaneswar virus	Whitefly	-	1
2. Bhendi yellow vein Delhi virus	Whitefly	-	1
3. Bhendi yellow vein Haryana virus	Whitefly	-	1
5. Bhendi yellow vein India virus	Whitefly	-	1
6. Bhendi yellow vein Tirupathi virus	Whitefly	-	1
7. Bhendi yellow vein Trichy virus	Whitefly	-	1
8. Bhendi yellow vein mosaic virus	Whitefly	1	1
9. Cotton leaf curl Alabad Virus	Whitefly	-	1
10. Cotton leaf curl Bangalore virus	Whitefly	-	1
11. Cotton leaf curl Multon virus	Whitefly	1	-
12. Cotton leaf curl Gezira virus	Whitefly	1	-
13. Okra Enation leaf curl virus	Whitefly	-	1
14. Okra leaf curl India virus	Whitefly	-	1
15. Okra leaf curl virus	Whitefly	1	-
16. Okra leaf curl Mali virus	Whitefly	1	-
17. Okra mottle virus	Whitefly	1	-
18. Okra yellow crinkle virus	Whitefly	1	-
19. Okra yellow mottle Iguala virus	Whitefly	1	-
20. Okra yellow mosaic Mexico virus	Whitefly	1	-
21. Okra yellow vein mosaic virus	Whitefly	1	1
22. Radish leaf curl virus	Whitefly	-	1
23. Sida micrantha mosaic virus	Whitefly	1	-
24. Tomato leaf curl Cameroon virus	Whitefly	1	-
25. Tomato leaf curl New Delhi virus	Whitefly	-	1
26. Okra yellow mottle Iguala virus	Whitefly	1	-
27. Okra yellow mosaic Mexico virus	Whitefly	1	-
B. *Ilarvirus*	Pollen	-	1
Tobacco streak virus	Pollen	-	1
C. *Nepovirus*	Nematode	1	-
Tobacco ring spot virus	Nematode	1	-

Virus group/Genus	Transmission/ Vector	No. Virus Species	
		World	India
D. *Potyvirus*	Aphid	1	-
Turnip mosaic virus	Aphid	1	-
E. *Tospovirus*	Thrips	-	1
Groundnut bud necrosis virus	Thips	-	1
F. *Tymovirus*	Beetle	1	-
Okra mosaic virus	Beetle	1	-
	Sub-Total	**17**	**19**
	Total	**36**	

Impact of Viruses in Okra

Yellow vein mosaic is most devastating disease of okra causing >60 per cent yield loss, Okra Enation Leaf Curl Virus serious in North India. Recently, TSV an ilarvirus is an emerging problem for okra production, Begomoviruses have high recombination rate and the presence of B-biotype whitefly were contributing to epidemics of begomoviruses in okra. Host genetic resistance to viruses is one of the most practical, economical and environmentally secures strategies for reducing yield losses in okra

Okra Yellow Vein Mosaic Virus

The disease was first reported in Bombay by Kulkarni in 1924, Uppal *et al.* (1940) established viral nature of disease and gave the name "Yellow Vein Mosaic", classified in the Family-Tymovirideae of Geminivirus, causes yield loss of 50 to 94 per cent, Distribution in Subtropical – June to Sept, in tropical areas (Feburary to June). The wild taxa of okra (*Abelmoschus* species) for identification of YVMV resistance species were found resistant to YVMV disease. The eight *Abelmoschus* species occur in India, among these, *A. esculentus* is the only known cultivated species. *A. moschatus* occur as wild species and is also cultivated for its aromatic seeds, while the rest six are truly wild types. The wild species occupy diverse habitats. The species *A. ficulneus* and *A. tuberculatus* spread over semi arid tracts of north and northwestern; *A. crinitus, A. manihot* and *A. tetraphyllus* and *A. pungens* are found in Tarai ranges and foot hills of Himalayas; *A. angulosus, A. moschatus, A. tetraphyllus* are widely distributed throughout western and eastern ghats and also peninsular tracts of south while *A. crinitus* and *A. pungens* are distributed in north eastern region of India. Assessions resistance to Yellow Vein Mosaic Virus (YVMV) in *A. manihot, A. angulosus, A. crinitus* and few land races of *A. tetraphyllus* (Singh *et al.*, 2007).

Okra Yellow Vein Mosaic Virus Disease Characterization

Homogenous inter-veinal network of yellow veins enclosing islands of green tissues. Initially infected leaves exhibit only yellowing of the veins and vein lets. In extreme cases the infected leaf becomes totally yellow or cream colored, Plants infected at the early stages remain stunted. The fruits of the infected plants exhibit pale yellow color, deformed, small and tough in texture. Disease adversely affects the both the quality of fruits and yield, the total loss has been estimated up to 20-30 per cent, if the pathogens are allowed to develop, this loss may increase up to 80-90 per cent. If the plants are infected within 20 days after germination, their growth is retarded, few leaves and fruits will form and loss may be about 94 per cent. The extent of damage declines with delay in infection of the plants. Plants infected 50 and 65 days after germination suffer a loss of 84 and 49 per cent respectively.

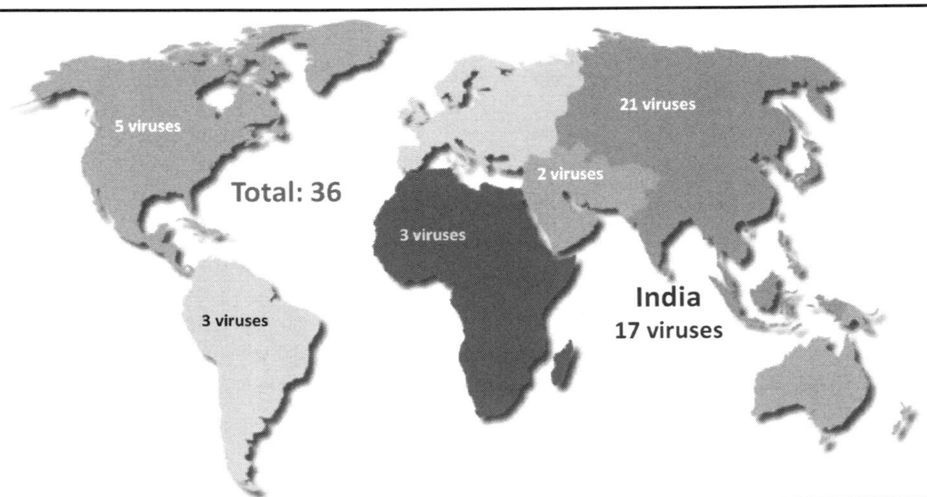

Begomoviruses	Ilarvirus	Nepovirus	Potyvirus	Tynovirus	Tospovirus
31	1-TSV	1-TbRSV	1-TuMV	1-OMV	1-GBNV

Figure 3.1. Global Distribution of Okra Viruses (Venkataramanappa *et al.*, 2011, 2012).

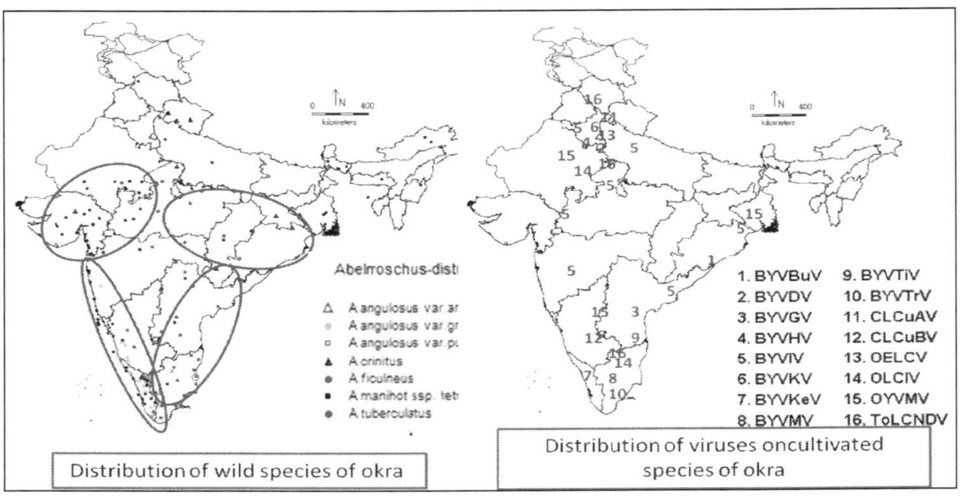

Figure 3.2. Distribution of Okra Wild Species and Begomoviruses.

Table 3.3: Inheritance of Resistance to Yellow Vein Mosaic (Venkataramanappa *et al.*, 2011, 2012)

Wild Source	*Varieties*	*Alleles/genes*
IC1542	Pusa Sawani	Recessive alleles
A. manihot subsp. *manihot*	Punjab Padmini, Punjab 7	Two Complementary dominant genes
A. manihot subsp. *manihot*	Parbhani Kranti, Pusa A 4	Single dominant gene
A. manihot subsp. *tetraphyllus*	Arka Abhay, Arka Anamika	Two Complementary dominant genes
A. Manihot subsp. *manihot*	Hisar Unnit	Two Complementary dominant genes

A. tetraphyllus ***A. tuberculatus*** ***A. moschatus***

A. angulosus ***A. ficulneus*** ***A. caillei***

Figure 3.3.Wild Species in okra as YVMV Resistance.

Breeding Methods for YVMV

Introduction: Process of introducing varieties from different centers of origin.

Selection: Superior genotypes having different superior morphological characters

Mutation: Developing mutants using different physical or chemical mutagen.

Backcross: Crossing of F_1 hybrid with either of the parent. It is followed to transfer disease resistance to the progeny.

Interspecific hybridization: Crossing of individuals of different species of same genus or different genus of same family.

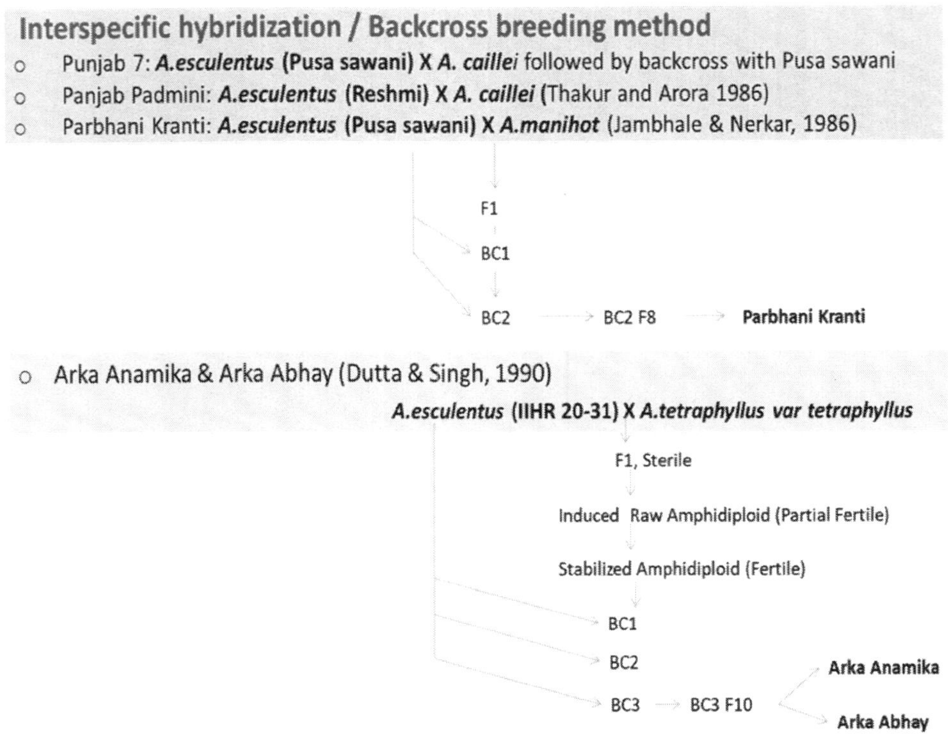

Figure 3.4. Breeding for YVMV Resistance.

Resistance Breeding involves following efficient screening techniques:

A. Efficient Screening Technique

1. Screening in hot spots/regions
2. Artificial Screening through whitefly
3. Screening through grafting/Agro inoculation

B. Resistance evaluation through Sensitive Assay

1. Visual symptoms
2. ELISA/Nucleic acid probe
3. PCR assay

The effect of okra yellow vein mosaic virus infection on PS II efficiency and thylakoid membrane protein changes in field grown bhendi leaves by the degree of virus infection, it was determined by means of the ratio of variable to maximum chlorophyll fluorescence (F_v/F_m). Changes in photosynthetic pigments, soluble proteins, nitrate reductase, photosynthetic activities and thylakoid membrane proteins in okra after okra yellow vein mosaic virus infection resulted in the reduction of photosynthetic pigments, soluble proteins, ribulose-1, 5-bisphosphate carboxylase activity and nitrate reductase activity. Virus infection caused marked inhibition of PS II activity. The artificial electron donors, diphenyl carbazide and hydroxylamine significantly restored the loss of PS II activity in infected leaves. The marked loss of PS II activity in infected leaves may be due to the loss of 47, 33, 28–25, 17 and 15 kDa polypeptides. It is concluded that the yellow vein mosaic virus infection inactivates the donor side of PS II. Virus coat protein seems to accumulate in the membranes of chloroplasts and thylakoid of the infected plants The physiological response of virus-infected bhendi leaves, by comparing the biochemical and physiological parameters of virus infected plants with healthy plants. It inhibit the electron transport activities in PS II, Chlorophyll, soluble protein and starch analysis, nitrate reductase activity, chlorophyll fluorescence, extraction and assay of RuBPCase activity, isolation of thylakoid membranes and DCPIP photo reduction, the total chlorophyll and carotenoids concentrations in infected leaves were significantly reduced by 64 per cent and 62 per cent, respectively. The content of soluble proteins was reduced markedly in infected leaves. The loss of leaf soluble protein in infected leaves would be partially accounted for damaged chloroplasts or could be the result of inhibition of protein synthesis. The decrease in photosynthetic pigments, nitrate reductase, photosynthetic electron transport activities and thylakoid membrane proteins in the infected leaves was due to YVMV infection induced rapid senescence in bhendi leaves (Palanisamy *et al.*, 2009). The resistant germplasm lines were observed in wild taxa of *A. tetraphyllus* followed by *A. callei* which could be used for developing resistant varieties. Some of the germplasm lines from wild and cultivated species of okra were found free from infection of YVMV in both the seasons, these lines were only categorized as immune after artificial grafting and serological confirmation. *A. manihot* ssp. *manihot* accession has been used in the development of YVMV resistant varieties (Sharma and Dhillon, 1983). *A. caillei* and *A. tetraphyllus* have been utilized in the improvement of convential varieties *i.e.* Punjab Padmini, P-7 and Parbhani Kranti, Arka Anamika and Arka Abhay. The resistant germplasm lines of wild taxa of *A. tetraphyllus* followed by *A. caillei* which could be used for the development of YVMV resistance varieties.

Breeding Achievements in Okra

Introduction	*A. manihot* ssp. *manihot* from Africa
	Perkins Long
	Clemsons Spineless
Selection	Pusa Sawani - Selection from Bihar local
	Pusa Makhmali – Pure line selection from West Bengal
	Co-1 from Red Wonder
Mutation Breeding	MDU-1 using gamma rays
	Punjab-8 using EMS
Hybridization	Most common method
Intra-specific hybridization	Pusa Sawani (1962), Sel 2
Inter-specific hybridization	Parbhani Kranti

Conclusion and Future Prospects

Identification of stable sources of resistance to the specific virus strains, introgression of resistance genes from wild species. Identification and isolation of specific virus resistant lines and divulging the genetics of YVMV inheritance and identification of resistance genes in different backgrounds. Breeding for YVMV disease resistance and high yielding lines from the available resistant source by hybridization programme, back cross breeding, biotechnology and other breeding methods plays important role. In recent past, the lacks of stability in YVMV resistance and there is a frequent break down.Therefore, there is a need to breed variety and hybrids for YVMV a resistant in coming days

REFERENCES

Arora D, Jindal SK, Singh K (2008) Genetics of resistance to yellow vein mosaic virus in inter-varietal crosses of okra (*Abelmoschus esculentus* L. Moench). *SABRAO J Breeding and Genet* 40(2): 93-103.

Deshmukh ND, Jadhav BP, Halakude IS, Rajput JC (2011) Identification of new resistant sources for yellow vein mosaic virus disease of okra (*Abelmoschus esculentus* L.). Veg Sci 38(1): 79-81.

Mohammad Ali, Hossain MZ, Sarker, NC. (2000) Inheritance of yellow vein mosaic virus (YVMV) tolerance in a cultivar of okra (*Abelmoschus esculentus* (L.) Moench). Euphytica 111: 205-209.

NHB (2017) Horticulture Crop Statistics of Karnataka state at a glance, published by Directorate of Horticulture, Lalbagh, Bangalore, pp.9

Palaniswamy P, Michael PI, Krishnaswamy M (2009) Physiological response of yellow vein mosaic virus infected bhendi (*Abelmoschus esculentus* L.) leaves. Physiol Mol Plant Path 74(1): 129-133.

Samarajeewa PK and Rathnayaka Rmusk (2004) Disease resistance and genetic variation of wild relatives of Okra (*Abelmoschus esculentus* L.). Annals of the Shri Lanka Department of Agriculture, 6(42): 167-176.

Shastry KSM, Singh SJ (1974) Effect of yellow vein mosaic virus infection on growth and yield of okra crop. Ind Phytopathol 27: 294-297.

Singh HB, Joshi BS, Khanna PP, Gupta PS (1962) Breeding for field resistance to yellow vein mosaic in bhindi. Indian J Genet 22(2): 137–144.

Singh M and M. R. Thakur, (1979) Nature of resistance to yellow vein mosaic virus in Abelmoschus manihot ssp. manihot. Current Sci 48(4): 164–165.

Singh B, Rai M, Kalloo G, Satpathy S, Pandey KK (2007) Wild taxa of okra (*Abelmoschus Species*): Reservoir of genes for resistance to biotic stresses. Acta Hort 20(3): 323-327.

Solankey SS, Shirin A, Randhir K, Verma RB, Kumar S. (2014) Seasonal response of okra (*Abelmoschus esculentus* L.) genotypes for okra yellow vein mosaic virus incidence. African J Biotechnol 13(12): 1336-1342.

Taware SD, Thengane, RJ, Harke SN, Taware AS (2010) Genome sequence analysis of DNA-A from okra yellow vein mosaic virus isolates. Int J Current Res 8: 96-107.

Thakur MR (1976) Inheritance of resistance to Yellow Vein Mosaic (YVM) in a cross of okra species, *Abelmoschus esculentus, A. manihot* ssp. *manihot*. *SABRAO J Breeding and Genet* 8(1): 69–73.

Vadivukarasi T, Girish KR, Usha R (2006) Sequence and recombination analyses of the geminivirus replication initiator protein. J Biosci 32: 17–29.

Venkataravanappa V, Lakshminarayana Reddy CN, Krishna Reddy M (2013) Begomovirus characterization, and development of phenotypic and DNA-based diagnostics for screening of okra genotype resistance against Bhendi yellow vein mosaic virus. 3 Biotech 3: 461-470.

Venkataravanappa V, Reddy CNL, Jalali S, Krishna Reddy M (2012) Molecular characterization of distinct bipartite begomovirus infecting bhendi (*Abelmoschus esculentus* L.) in India. Virus Genes 44(3): 522-535.

Arora D, Jindal SK, Singh K (2008) Genetics of resistance to yellow vein mosaic virus in inter-varietal crosses of okra (Abelmoschus esculentus L. Moench.) *SABRAO J Breeding and Genet* 40(2): 93-103.

Deshmukh ND, Jadhav BP, Halakude IS, Rajput JC (2011) Identification of new resistant sources for yellow vein mosaic virus disease of okra (*Abelmoschus esculentus* L.) Veg Sci 38(1): 79-81.

Palaniswamy P, Michael PI, Krishnaswamy M (2009) Physiological response of yellow vein mosaic virus infected bhendi (*Abelmoschus esculentus* L.) leaves. *Physiol Mol Plant Path* 74(1): 129-133.

Singh B, Rai M, Kalloo G, Satpathy S, Pandey KK (2007) Wild taxa of okra (*Abelmoschus Species*): Reservoir of genes for resistance to biotic stresses. *Acta Hort* 20(3): 323-327.

Solankey SS, Shirin A, Randhir K, Verma RB Kumar S (2014) Seasonal response of okra (*Abelmoschus esculentus* L.) genotypes for okra yellow vein mosaic virus incidence. *African J Biotechnol* 13(12): 1336-1342.

Chapter 4

Breeding of Vegetables for Nutritional Quality

Abhilash Singh, Shubham Singh, Sandeep Yadav and Ramavath Ramesh Babu

ABSTRACT

Vegetables are fundamental components of human diets. They not only are important sources of essential vitamins and minerals, but also contain a wide variety of secondary metabolites important to human health (Chiu et al., 2010). The phytochemical impart health benefits and provide protection against certain degenerative diseases like blindness, cancer, diabetic. Vitamin-A deficiency is quite problem in developing country like Africa and South East Asia and 140 million preschool-aged children and 1.2 to 1.3 million children affected every year (UN –SCN 2004), hence these problems can be combat by developing colour rich nutritional vegetable varieties. Pigments in plants have four major classes are chlorophylls, carotenoids, flavonoids, and betalins. Colour rich vegetables attract visitors and eaters (Chen, 2015). Among the carotenoids, B-carotene (pro-vitamins A) is precursor of Vit-A and orange and yellow colour vegetables are like orange carrot (3000-6000µg/100g), orange cauliflower (Or gene linked SCAR marker) having (80-320µg/100gm), yellow cucumber (22-48µg/100gm), yellow potato (200-500µg/100gm), orange tomato (3000-4500µg/100gm) are highly revered as a "good for the eyes". Lycopene is the antioxidant red pigment in vegetables such as tomato and watermelon (Faree et al., 2010). Flavonoids, a group of secondary metabolites belonging to the class of phenyl-propanoids have the widest colour range, from pale-yellow to blue (Tanaka et al., 2016). Anthocyanin, naturally occurring pigment with high antioxidant responsible for red, blue, purple colour in vegetables like eggplants, onion, red cabbage, purple cabbage (Zhang et al., 2016). In the recent year certain improved anthocyanin concentration (100gm) in vegetables crops like purple carrot (200-350mg), purple potato (17-20mg), red flesh potato (20-38mg), red onion (25-40mg), red cabbage (200-3320mg), purple tomato (20-60mg). In research purpose various mutants and transgenic gene identified like Or (cauliflower) (Zhou et al., 2008) and MYB (red cabage and purple cauliflower), Aft, Abg, atv (tomato) respectively. Colorful vegetables can be breed be by the use of different breeding methodology like Selection, Hybridization, Somaclonal variation and transgenic breeding but now days transgenic breeding is the main approach to enhance the colour of the vegetables. Till date numerous transgenic lines have been developed mainly tomato, egg plant, sweet pepper, potato and other vegetables crops.

Keywords: *Vegetable breeding, Nutraceutical, Genetic engineering and Quality traits.*

Introduction

Vegetables are considered essential for well-balanced diets since they supply vitamins, minerals, dietary fiber, and phytochemicals. A world vegetable survey showed that 402 vegetable crops are cultivated worldwide, representing 69 families and 230 genera. Leafy vegetables of which the leaves or young leafy shoots are consumed were the most often utilized (53 per cent of the total), followed by vegetable fruits (15 per cent), and vegetables with below ground edible organs comprised 17 per cent. Each vegetable group contains a unique combination and amount of these phyto-nutriceuticals, which distinguishes them from other groups and vegetables within their own group. Some phytochemicals of vegetables are strong antioxidants and are thought to reduce the risk of chronic disease by protecting against free radical damage, by modifying metabolic activation and detoxification of carcinogens, or even by influencing processes that alter the course of tumor cells.

Nutrition is both a quantity and a quality issue, and vegetables in all their many forms ensure an adequate intake of most vitamins and nutrients, dietary fibers, and phytochemicals which can bring a much-needed measure of balance back to diets contributing to solve many of these nutrition problems. The promotion of healthy vegetable products has coincided with a surging consumer interested in the healthy functionality of food. It has been estimated to cause about 31 per cent of ischaemic heart disease and 11 per cent of stroke worldwide. According to the 2007 World Health Report unbalanced diets with low vegetable intake and low consumption of complex carbohydrates and dietary fiber are estimated to cause some 2.7 million deaths each year, and were among the top 10 risk factors contributing to mortality.

Consumer's interest in whole foods with enhanced nutritional qualities is at an all-time high, and more consumers are choosing foods on the basis of their healthy benefits. For example the *Apiaceae* family (*e.g.* celery, parsley, carrot) is rich in flavonoids, carotenoids, vitamin C, and vitamin E. Celery and parsley for example are among the best vegetables sources for the flavonoid apigenin and vitamin E, and carrots have an unique combination of three flavonoids: kaempferol, quercetin, and luteolin. In carrot, overall carotenoid levels, have increased dramatically in the past four decades through traditional breeding to reach levels of 1000 ppm carotenoids, on a fresh weight basis. The *Asteraceae* or *Compositae* family (*e.g.* lettuce, chicory) is rich in conjugated quercetin, flavonoids, and tocopherols and observed sizeable variations in flavonol content were also observed with lettuce cultivars. The red color of this lettuce is due to high levels of anthocyanins, which like quercetin, are products of the phenylpropanoid pathway.

Malnutrition

The World Health Organization defined the "Malnutrition" which refers to deficiencies, excesses or imbalances in a person's intake of energy and/or nutrients. The term malnutrition covers 2 broad groups of conditions. One is 'undernutrition'—which includes stunting (low height for age), wasting (low weight for height), underweight (low weight for age) and micronutrient deficiencies or insufficiencies (a lack of important vitamins and minerals). The other is overweight, obesity and diet-related non-communicable diseases (such as heart disease, stroke, diabetes and cancer)".

About 3 billion people in the world are malnourished due to imbalanced diets. Under consumption of vegetables and fruits is among the top 10 risk factors leading to micronutrient malnutrition and is associated with the prevalence of chronic diseases. More than 70 per cent of malnourished children live in Asia. Deficiency of vitamin A and iron can result in severe anemia, impaired intellectual development, blindness, and even death. Vegetables contain a range of macro and micro-nutrients, including pro-vitamin A, iron and zinc, which contribute to the prevention of malnutrition disorders. Small variation in maternal diets, particularly reduction in micronutrient content, can have a significant impact on fetal growth and development. Pregnant women, in particular, benefit from good vegetable nutrition, particularly during later pregnancy, and lactation. The interplay of the different micronutrients and antioxidants found in vegetables has important health impacts, explaining for instance the higher birth weight of children in India, when mothers consume higher rates of green leafy vegetables and fruits during pregnancy. Nutrition is both a quantity and a quality issue, and vegetables in all their many forms ensure an adequate intake of most vitamins and nutrients, dietary fibers, and phytochemicals, which can bring much needed balanced diets thereby contributing to solve many of these nutritional problems.

Nutritional Quality

Phyto-neutriceuticals are all the chemical compounds derived from plants that have health promotion properties. Most of them are found in relatively small quantities in vegetables. However, when consumed in sufficient quantities, they contribute significantly toward protecting living cells against chronic diseases. The most important phyto-neutriceuticals in vegetables that have biological activity against chronic diseases are: vitamins, minerals, dietary fiber, organosulfur compounds (glucosinolates and thiosulfides) and flavonoids.

Protein

Protein is a macronutrient, and body needs protein in large amounts. It is essential for the maintenance and building of body tissues and muscle. 80 g protein per day needed for an adult.

Protein rich vegetables are Lima bean (7.9 g/100g), Pea (7.2 g/100g), Cowpea (4.3 g/100g), Spinach (4 g/100g).

Vitamins

The term vitamin is derived from the words vital and amine, because vitamins are required for life and were originally thought to be amines. Although not all vitamins are amines. They are any of various fat-soluble or water-soluble organic substances essential in minute amounts for normal growth and activity of the body and obtained naturally from plant and animal foods. A healthy body needs vitamins. All vegetables are sources of vitamins and consequently have health benefits. In this section, we will highlight and discuss only the vitamins with a high prevalence of inadequate intake in worldwide diets such as vitamins C, A, and E and folates.

Vitamin A

Vitamin A is a major public health problem in much of the developing world. The most abundant carotenoids in vegetables are α-carotene, β-carotene, lycopene, lutein, zeaxanthin, and β-cryptoxanthin. These carotenoids account for more than 90 per cent of the carotenoids present in the human diet. Carotenoids with molecules containing oxygen, such as lutein, zeaxanthin and β-cryptoxanthin, are known as xanthophylls. The un-oxygenated carotenoids such as α-carotene, β-carotene, and lycopene are known as carotenes. The most common carotenes are β-carotene and lycopene. Lutein is the most abundant xanthophyll. Only three of the carotenoids (α-carotene, β-carotene, and β-cryptoxanthin) can be converted into the provitamin A (retinol), while lycopene, lutein, and zeaxanthin have no vitamin A activity. β-carotene is the carotenoid with the most provitamin A activity, It is the more common form, and it is also one of the most thoroughly studied. The bioavailability of β-carotene from vegetables in the human diet is limited. Studies have estimated that β-carotene bioavailability is 22 to 24 per cent from broccoli, 19 to 34 per cent from carrots, and 3 to 6 per cent from leafy vegetables. Several factors have been shown to influence carotenoids' bioavailability in humans, the most prominent of which is the food matrix.

Vitamin C

Vitamin C is an essential vitamin to humans that lack the ability to synthesize this vitamin because they are deficient in the enzyme L-gulonolactone oxidase, an enzyme involved in the biosynthesis of vitamin C via the glucuronic acid pathway. The biological function of vitamin C is based on its ability to donate electrons, which provide intra- and extracellular reducing power for a variety of biochemical reactions. The reducing power of vitamin C is capable of neutralizing most of the physiologically relevant reactive oxygen and nitrogen species in the human body. Several enzymes involved in biosynthesis of collagen, carnitine and neurotransmiters require vitamin C. As early as the 15th century, vitamin C was referred to as the "antiscorbutic factor," since it helped prevent the potentially fatal disease called scurvy. This disease was first discovered in Portuguese sailors, whose sea voyages left them far away from natural surroundings for long periods of time. Their body stores of vitamin C fell below 300 mg, and their gums and skin lost the protective effects of vitamin C.

Vitamin C is a water-soluble vitamin and is highly sensitive to water, air, and temperature. About 25 per cent of the vitamin C in vegetables can be lost simply by blanching (boiling or steaming the food for a few minutes). This same degree of loss occurs in the freezing and unthawing of vegetables. Cooking of vegetables for longer periods of time (10-20 minutes) can result in a loss of over one-half the total vitamin C content. When vegetables are canned and then reheated, only one third of the original vitamin C content may be left. For these reasons, consumption of vitamin C-rich vegetables in their fresh, raw form is the best way to maximize vitamin C intake. Vitamin C has significant interactions with several key minerals in the body. Supplemental intake of vitamin C at gram level doses can interfere with copper metabolism.

Vitamin E

Vitamin E is a generic term for all tocopherols and their derivatives having biological activity. In nature, it is present under eight different forms, four tocopherols (α-, β-, γ-, δ) and four tocotrienols (α-, β-, γ-, δ), of which α-tocopherol is the most predominant in nature and most bioactive form. The α, β, γ, and δ isoforms of tocopherol have relative potency of 100 per cent, 50 per cent, 10 per cent and 3 per cent, respectively. Since tocopherols are of non-polar nature, their main function lies in their action in the hydrophobic environment of cell and organelle membranes protecting these structures from free radicals and also by stabilizing them. Tocopherols induce a protective effect against oxidative stress linked to metabolic syndrome as well as other sources. They are also essential for normal neurological function. As an antioxidant, α-tocopherol is able to prevent free-radical mediated tissue damage and thus to prevent or delay the development of degenerative and inflammatory diseases.

Vitamin E deficiency is believed to be associated with the pathogenesis of cardiovascular disease including low-density lipoprotein oxidation, cytokine production, production of lipid mediators, platelet function, and smooth muscle cell proliferation, as well as interaction of the endothelium with immune and inflammatory cells. The current recommended dietary allowance for vitamin E intake from natural foods is about 10-15 mg/day.

Minerals

The vegetables we eat absorbed plenty of minerals as they were growing. When we eat those vegetables, we absorb the minerals as we digest food. All vegetables are sources of minerals and consequently have health benefits. A healthy body needs plenty of minerals. Minerals can be divided into two groups. There are some that we need only in tiny quantities (trace) and others that we need much more of (major). Major minerals, macronutrients, with a high prevalence of inadequate intake in worldwide diets include calcium, magnesium, and potassium. Trace minerals, micronutrients, identified as nutrients with great health implications and benefits, usually with a high prevalence of inadequate intake, include iron, zinc and selenium.

Calcium is an important component of a healthy diet and a mineral necessary for life. It is essential for strong bones and teeth. Calcium plays an important role in building stronger, denser bones early in life and keeping bones strong and healthy later in life.Approximately 99 per cent of the body's calcium is stored in the bones and teeth.

Magnesium helps our body to release energy from food. Magnesium is needed for bone, protein, making new cells, activating B vitamins, relaxing nerves and muscles, clotting blood, and in energy production. Insulin secretion and function also require magnesium. Magnesium also assists in the absorption of calcium, vitamin C and potassium. Deficiency may result in fatigue, nervousness, insomnia, heart problems.

Potassium is the eighth most common element by mass (0.2 per cent) in the human body, so that a 60 kg adult contains a total of about 120 grams of potassium. Potassium ions are necessary for the function of all living cells. It is essential for the body's growth and maintenance. Potassium cat-ions are important in neuron (brain and nerve) function and in influencing osmotic balance between cells and the interstitial fluid, with their distribution mediated in all animals by the so-called Na+/K+-ATPase pump.

Iron plays an important role in human health, forming complexes with molecular oxygen in hemoglobin and myoglobin. These two compounds are common oxygen transport proteins in vertebrates. Iron is also the metal used at the active site of many important redox enzymes dealing with cellular respiration and oxidation and reduction in plants and animals. Eradicating iron deficiency can improve national productivity levels by as much as 20 per cent. Most at risk of iron deficiency are infants, adolescent girls and pregnant women. Pregnant women are at special risk of low iron levels and are often advised to supplement their iron intake. Iron deficiency in infants can result in impaired learning ability and behavioral problems. It can also affect the immune system and cause weakness and fatigue. Vitamin C promotes the absorption of non-heme iron. It is necessary to eat foods rich in vitamin C at the same time we eat the food containing iron.

Zinc is an essential mineral of exceptional biologic and public health importance. Zinc is found in nearly 100 specific enzymes, serves as structural ions in transcription factors and is stored and transferred in metallothioneins. It is the second most abundant transition metal in organisms after iron, and it is the only metal that appears in all enzyme classes. Zinc supplements help prevent disease and reduce mortality, especially among children with low birth weight or stunted growth. However, zinc supplements should not be administered alone, since many in the developing world have several deficiencies, and zinc interacts with other micronutrients.

Dietary Fiber

Dietary fiber is a heterogeneous mixture of polysaccharide carbohydrates, such as cellulose, hemicellulose, β-glucans, pectins, mucilages and gums, plus noncarbohydrates, such as lignin, which are indigestible in the small intestine. Dietary fiber is classified into water-soluble and water-insoluble fiber.

Antioxidants

Lycopene

Lycopene was found to be more bioavailable from the thermally processed tomato products like tomato paste than from fresh tomato due to heat treatment and the presence of higher oil content in the paste. Lycopene exists in food primarily in the *trans* sterisomeric configuration, however, cooking and processing help convert *trans*-lycopene to *cis*-lycopene, which is more readily absorbed through the intestinal wall into the plasma.

Capsaicinoids

Belonging to two groups, capsaicin and dihydrocapsaicin, have been identified in pepper. Capsaicin was discovered in 1846, and its structure, as an acid amide, was elucidated by Nelson. Capsaicin constitutes about 70 per cent of the pungent flavour in hot pepper, while its analogue dihydrocapsaicin represents 30 per cent. The two groups differ in the presence or absence of double bonds in the fatty acid side-chain, and within each group, they differ in the length and branching point in the fatty acid side-chain. Significant variations in the profile of capsaicinoids are found between and within pepper species, ranging from about 220 ppm (3,400 Scoville Heat Units, SHU) in *Capsicum annum* to 20,000 ppm (320,000 SHU) in *Capsicum chinense.*

Breeding Approaches

1. Evaluation of germplasm and selection
2. Pure line Selection
3. Pedigree Method
4. Hybridization
5. Interspecific hybridization
6. Somaclonal variations
7. Mutagenesis
8. Genetic engineering

Evaluation of Germplasm and Selection

A total of 49 accessions of tomato germplasm from 24 countries on 4 continents, provided by the Instituto de Conservacio´n y Mejora de la Agrodiversidad Valenciana (COMAV) genebank, were evaluated. Fourteen accessions of common tomato types (*Solanum lycopersicum* L.), 28 cherry type tomatoes (*S. lycopersicum* var. *cerasiforme* L.) and 7 accessions of small fruit tomato related species (*Solanum pimpinellifolium* L.), which represent a wide diversity of fruit shapes and colours, Fourteen accessions of the cherry type and two of the common tomato type were selected for their high and balanced nutritional properties, causing them to be of great interest for direct human consumption (especially BGV008057, BGV006863 and BGV008060). Furthermore, BGV008365 and BGV012627 (cherry types with over 1.5 times the normal average ascorbic acid content) as well as BGV008166 (*Solanum pimpinellifolium*) accession which presented more than nine times the normal average lycopene content) would be of interest as donor parents for breeding programmes to increase the nutrition properties of commercial varieties (Adalid *et al.*, 2009).

Pure Line Selection

A pure line is progeny of single homozygous plant of self pollinated species. Therefore, all the plants in a pure lines of the crops Amaranth- Arka Arunima (Rich in Ca, Fe), Chilli- Arka Abhir (Capsainthin) would be most appropriate for cultivation.

Pedigree Method

In this method selected parents are crossed in crossing block to produce F_1 and F_2 generation. The selections then made for phenotypically superior plant. It is useful by finding out if two individuals are related by descent, have a common parent in their ancestors and they have some genes in common like in onion variety Arka Swadista (high TSS).

Hybridization

Hybridization includes crossing between two genetically diverse genotypes. Monogenic nutritional characters can be brought into cultivated types by following back cross method of breeding. Recurrent selection is followed to increase desirable gene frequencies and to break undesirable linkages. Potato varieties Kufri Chipsona 1 and K. Chipsona 2 having high dry matter and low sugar. In watermelon, Mateera AHW 19, Mateera AHW 5 having High TSS. In cowpea, Bidhan Barbati 1 and 2 having high protein. In Spinach, Arka Anupama having high vitamin A, C, Carotene and Fe.

Interspecific Hybridization

Wild relatives of a crop includes its primitive forms (from where they have originated), related weedy species and other species in the same genus, which are not under cultivation. Wild relatives of a crop includes its primitive forms (from where they have originated), related weedy species and other species in the same genus, which are not under cultivation. There are several example for utilization of wild relative in quality improvement like, *Solanum esculentum* x *Solanum hirsutum* –high beta carotene, *Solanum esculentum* x *Solanum hirsutum* f.sp. *glabratum* – high beta carotene. There are the several wild relative for their utilization on quality improvement like tomato *S. hirsutum*, *S. pimpinellifolium*, *S. glabratum*, *S. cheesmanii*, *S. peruvianum* having rich in β carotene,vitamin-C, TSS, carotene, TSS, vitamin-C, anthocyanin respectively. In potato like *S. microdontum*, *S. vernei*, *S. phureja spp. phureja* are rich source of calcium, starch and protein respectively.

Somaclonal Variations

"Scarlet" variety of sweet potato developed through somaclonal variation is having high starch content. Tomato variety "DNAP-9" showed 20 per cent more soluble solids (Liu, 2010).

Mutation

The utilization of induced mutations for crop improvement is known as mutation breeding. In tomato, this has been successfully utilized to evolve S-12 from Sioux and Pusa Lal Meeruti from Improved Meeruti and Marutham has been evolved by treating seeds of CO-1 with EMS-1. In sweet-potato, the mutant "Port Blaco" from Port-Rico has revealed white flesh with high starch.

Genetic Engineering

Biotechnology is a new, and potentially powerful, tool that has been added by all the major seed corporations to their vegetable breeding research programs. It can augment and/or accelerate conventional cultivar development programs through time saved, better products, and more genetic uniformity, or achieve results not possible by conventional breeding. Transgenic crops, commonly referred to as genetically modified (GM) crops, enable plant breeders to bring favorable genes, often previously inaccessible, into already elite cultivars, improving their value considerably and offering unique opportunities for controlling insects, viruses and other pathogens, as well as nutritional quality. To enhance the carotenoid content and profile of tomato fruit, Romer *et al.* (2000) have produced transgenic lines containing a bacterial carotenoid gene (*crtI*) encoding the enzyme phytoene desaturase. Lycopene is transformed into β-carotene by the action of lycopene beta-cyclase (beta-Lcy). Rosati *et al.* (2000) introduced, via *Agrobacterium*-mediated transformation, DNA constructs aimed at upregulating (OE construct) or downregulating (AS construct) the expression of the beta-Lcy gene in a fruit-specific fashion. which converts phytoene into lycopene. Expression of this gene in transgenic tomato plants of the cultivar "Ailsa Cray" did not elevate total carotenoid levels. However, the β-carotene content increased about threefold, up to 45 per cent of the total carotenoid content.

Improve Protein Quantity and Quality

Introducing genes from other sources for proteins that have a favourable balance of essential amino acids *e.g.* introduction of gene for seed albumin protein from amaranth into potato. Increasing the production of specific amino acids such as methionine. Three genes, encoding phytoene synthase (*CrtB*), phytoene desaturase (*CrtI*) and lycopene beta-cyclase (*CrtY*) from *Erwinia* in potato to produce β-carotene. Over expression of *AmA1* results in increased tuber protein and amino Acids There was a 35–60 per cent increase in total protein content in moderately expressed transgenic events of all varieties. The balance of amino acid is a key determinant for nutritive proteins because plant proteins typically do not provide the optimum amino acid ratios required for efficient protein synthesis in animals. In tomato *Anthocyanin fruit* (*Aft*) genotype is characterized by purple color in skin and outer pericarp of its fruits due to higher levels of anthocyanins—flavonoid metabolite *high pigment-1* (*hp-1*) mutation, known to increase flavonoids in tomato fruits an anthocyanin mutant (*ant1*) with prominent purple coloration. The inheritance of the *ant1* mutant followed a typical Mendelian pattern for a recessive trait controlled by a single gene.

Introgression of *Or* Gene in Indian Cauliflower

Curd colour phenotype observed in F_2 population of cauliflower process (18-19 white colour x 1227 inbreed line). *Or* gene is control by single dominant gene with modifier. Marker assisted introgression of '*Or*' gene for enhancing β-carotene in cauliflower through marker assisted backcrossing. Pusa Kesari is first carotene rich variety released for sowing in October–January. Curds compact, orange coloured, very attractive with semi-self blanching growth habit and weighing about 1.250 kg with 9-10 ppm β-carotene.

Table 4.1. Varieties that are Developed through Genetic Engineering/Biotechnology

Crop	*Variety*	*Character Improved*
Tomato	Pusa Red Plum, Hissar Arun, CO-3	Vitamin-C
	Pusa Rohini	β- Carotene
	Pusa Ruby	TSS
Chilli	Punjab Lal	Capsaicin
	Arka Abhir	Capsanthin (Paprika)
	KTPL-19	Capsanthin (Paprika)
Broccoli	Palam Vichitra	Anthocyanin
	Palam Kanchan	Carotene
Sem	Purple Sem	Anthocyanin
Onion	Pusa Riddhi	High Antioxidant
	Arka Swadista	High TSS
	Punjab-48	High TSS
Amaranthus	Pusa Lal Chaulai	Anthocyanin
Sweet Potato	Bhu Sona	β- carotene
	Bhu Krishna	Anthocyanin
	Pusa Sunehari	β- carotene
	Sree Vardhani	β- carotene
	Sree Ratna	β- carotene
	Kishan	High Starch
Topioca	Sree Visakham	β- carotene
Potato	Kufri Chipsona-1,2	High TSS and Dry matter
Muskmelon	Arka Jeet	High TSS
Watermelon	Arka Jyoti	High TSS
Carrot	Pusa Ashita	Anthocyanin (Black colour)
	Pusa Rudhira	Lycopene(Blood red colour)
	Pusa Meghali	Vitamin-A
	Pusa Nayanjyoti	β- Carotene
Beet root	Detroit Dark Red	Anthocyanin
Radish	Pusa Gulabi	Anthocyanin
	Pusa Jamuni	Anthocyanin

Crop	Variety	Character Improved
	Palam Hridya	Anthocyanin
	Pusa Vidhu	White coloured
	Pusa Sagarika	Purple flesh variety
Turnip	Pusa Kanchan	Carotene
Cauliflower	Pusa Beta Kesari	β- Carotene
Cabbage	Red Cabbage	Anthocyanin

Conclusion

Various phyto-nutriceuticals with antioxidant properties may work directly by quenching free radicals or indirectly by participating in cell-signaling pathways. Sensitive to redox balance, nutrients such as potassium contribute to blood pressure regulation. The dietary fiber content and type of different vegetables may also contribute to the overall health benefit, such as improving bowel transit, lowering cholesterol, helping manage blood glucose concentrations, and by transporting a significant amount of minerals and phytochemicals linked to the fibre matrix through the human gut. Phytonutriceuticals (vitamins, minerals, dietary fiber and phytochemicals), a great diversity of vegetables should be eaten to ensure that individual's diet includes a combination of phytonutriceuticals and to get all the health benefits. The availability of a large diversity of vegetables year round, and the knowledge of vegetable health benefits, has enabled consumers to include a variety of health-promoting phytonutriceuticals in their diets. Consequently, there is a good potential for increasing nutrient and phytochemical content and thus enriching the diet of the average consumer. Genetic improvement to increase levels of specific nutrients and phytochemicals is complex because there is often a large environmental effect when the component is present in tiny amounts. Success in vegetable breeding for higher vitamin, mineral and phytochemical content should consider not only substance concentration but also organic components in plants that can either reduce or increase bioavailability.

Research on the health benefits of vegetables, from a horticultural and breeding perspective, needs to focus on key areas in the near future such as:

1. To continue the evaluation of phytonutriceutical content among older versus newer major cultivars.
2. To identify the genetic mechanisms that regulate the synthesis of their key phytochemicals, such as the glucosinolates, thiosulfides and flavonoids, in order to develop cultivars rich in a variety of phytochemicals and in order to ensure that a mixture of phytochemicals enters into the human diet.
3. To study the potential change in the balance of these compounds (and their eventual synergisms or interactions).
4. To identify the optimum conditions for maintaining these phytochemicals after harvest and processing since studies have shown that the

bioavailability of some of the phytochemicals increases dramatically after storage and processing, while the bioavailability of others degrades rapidly.

Molecular tools will be useful for increasing quality, nutritional value, and yields. Transgenic crop cultivars should not be excluded as products capable of contributing to more nutritious and healthy vegetables. Biotechnology-derived vegetable crops will succeed if clear advantages and safety are demonstrated to both growers and consumers. Progress in breeding for human nutrition and health benefits will necessitate forming teams that besides plant breeders and molecular biologists also include nutritionists, health professionals and food scientists and processors.

REFERENCES

Abdelwahab M, Youssef S, Aly A, Elfiki S, Elenany N, Abbas M (1992) A simple calibration of a whole-body counter for the measurement of total body potassium in humans. Int J Radiat Applic Instrumen Part A Appl Radiat Isotopes **43**(10): 1285–1289.

American Heart Association (2000). Heart and stroke statistical update. American Heart Association. Dallas, Texas.

Atkinson J, Epand RF, Epand RM (2008) Tocopherols and tocotrienols in membranes: a critical review. Free Radic Biol Med 44(5): 739–764.

Bremner JD, McCaffery P (2008) The neurobiology of retinoic acid in affective disorders. Prog Neuropsychopharmacol Biol Psychiatry 32(2): 315-331.

Britton G, Khachik F (2009) Carotenoids in Food. In: Britton, G., Pfander, H. and Liaaen-Jensens, S. (eds.). Carotenoids. Birkhäuser, Basel. pp. 45-66.

Craig W, Beck L (1999) "Phytochemicals: Health Protective Effects," Canadian Journal of Dietetic Practice and Research 60: 78-84.

Cuddihy SL, Parker A, Harwood DT, Vissers MC, Winterbourn CC (2008) Ascorbate interacts with reduced glutathione to scavenge phenoxyl radicals in HL60 cells. Free Radic Biol Med 44(8): 1637-1644.

Dawson-Hughes B, Harris SS, Krall EA, Dallal GE (1997) Effect of calcium and vitamin D supplementation on bone density in men and women 65 years of age or older. N Engl J Med 337(10): 670–676.

Dias JS (2011) "World importance, marketing and trading of vegetables," Acta Horticulturae 921: 153-169.

Dias JS, Ryder E (2011) "World Vegetable Industry: Production, Breeding, Trends," Horticulture Review 38: 299-356.

Dreher ML (2001) Dietary fiber overview. In: Cho, S.S. and Dreher, M.L. (eds.). Handbook of dietary fiber. Marcel Dekker, New York.

Foyer CH, Noctor G (2011) Ascorbate and glutathione: the heart of the redox hub. Plant Physiol. 155(1): 2-18.

Frei B, Lawson S (2008) Vitamin C and cancer revisited. Proc. Natl. Acad. Sci. (USA) 105(32): 11037-11038.

Halpner AD, Handelman GJ, Belmont CA, Harris JM, Blumberg JB (1998) Protection by vitamin C of oxidant-induced loss of vitamin of vitamin E in rat hepatocytes. J Nutr Biochem 9: 355-359.

Hellen Keller (2010) International, "Homestead Food Production Model Contributes to Improved Household Food Security, Nutrition and Female Empowerment—Experience from Scaling- Up Programs in Asia (Bangladesh, Cambodia, Nepal and Philippines)," Nutrition Bulletin, 8: 1-8.

Herrera E, Jimenez R, Aruoma OI, Hercberg S, Sanchez-Garcia I, Fraga C (2009) Aspects of antioxidant foods and supplements in health and disease. Nutrition Reviews, 67: 140-144. doi: 10.1111/j.1753-4887.2009.00177.x

Hu J, Liu Y, Yu Y, Zhao TZ, Liu SD, Wang QQ (1991) Diet and cancer of the colon and rectum: a case-control study in China. Int J Epidemiol 20: 362-367.

Jacob RA (1995) The integrated antioxidant system. Nutr Res 15: 755-766.

Keatinge JDH, Waliyar F, Jammadass RH, Moustafa A, Andrade M, Drechsel P, d'A. Hughes J, Kardivel P, Luther K, (2010) "Re-Learning Old Lessons for the Future of Food: By Bread Alone No Longer—Diversifying Diets with Fruit and Vegetables" Crop Science, 50: 51-62.

Levine M, Rumsey S, Wang Y, Park J, Kwon O, Xu W, Amano N (1996) Vitamin C. In: Ziegler EE and Filer LJ (eds.). Present Knowledge in Nutrition. 7th Edition. ILSI Press, Washington DC. pp.146-149.

Liu S, Lee IM, Ajani U, Cole SR, Buring JE, Manson JE (2001) Intake of vegetables rich in carotenoids and risk of coronary heart disease in men: the physicians' health study. Int. J. Epidemiol. 30: 130-135.

Mullie P, Clarys P (2011) Association between cardiovascular disease risk factor knowledge and lifestyle. Food Nutr Sci 2: 1048-1053.

National Research Council (USA) (1989). Committee on Diet and Health, National Academy of Sciences. Diet and Health. National. Academies Press, Washington, DC.

Nelson DL, Cox MM (2005) Lehninger, Principles of Biochemistry. 4th Editon. WH Freeman and Company, New York.

Packer L, Fuchs J (eds.) (1997). Vitamin C in health and disease. Marcel Dekker, New York.

Peipins LA, Sandler RS (1994) Epidemiology of colorectal adenomas. Epidemiol Rev 16: 273-279.

Pfeiffer W, McClafferty B (2007) "HarvestPlus: Breeding crops for better nutrition," Crop Science 57: 88-105.

Rao S, Yajnik CS, Kanade A, Fall DHD, Margetts BM, Jackson AA (2001) Intake of micronutrient-rich foods in rural Indian mothers is associated with the size of their babies at birth: Pune maternal nutrition study. J. Nutr 131: 1217–1224.

Rao S, Yajnik CS, Kanade A, Fall DHD, Margetts BM, Jackson AA (2001) "Intake of micronutrient rich foods in rural Indian mothers is associated with the size of their babies at birth: Pune maternal nutrition study," Journal of Nutrition 131: 1217-1224.

Ravaglia G, Forti P, Lucicesare A, Pisacane N, Rietti E, Mangialasche F, Cecchetti R, Patterson C, Mecocci P (2008 Plasma tocopherols and risk of cognitive impairment in an elderly Italian cohort. Am J Clin Nutr 87(5): 1306-1313.

Romer S, Fraser PD, Kiano JW, Shipton CA, Misawa N, Schuch W, Bramley PM (2000). Elevation of the provitamin A content of transgenic tomato plants. Nat. Biotech. 18: 666–669.

Rosati C, Aquilani R, Dharmapuri S, Pallara P, Marusic C, Tavazza R, Bouvier F, Camara B, Giuliano G (2000) Metabolic engineering of beta-carotene and lycopene content in tomato fruit. Plant J 24: 413–419.

Southon S (2000) "Increased fruit and vegetable consumption within the EU: potential health benefits." Food Research International 33: 211-217. doi: 10.1016/S0963- 969(00)00036-3

Steinmetz KA, Kushi LH, Bostick R, Folisom AR, Potter JD (1994) Vegetable, fruit and colon cancer in the Iwoa women's health study. Am J Epidemiol 139: 1-15.

Traber MG, Atkinson J (2007) Vitamin E: antioxidant and nothing more. Free Radic Biol Med 43 (1): 4–15.

Trinidad TP, Mallillin AC, Loyola AS, Sagum RS, Encabo RR (2010) The potential health benefits of legumes as a good source of dietary fibre. Br J Nutr 103(4): 569-574.

UNICEF (2010) Vitamin and Mineral Deficiency. A Global Progress Report p 2.

Wargovich MJ (2000) "Anticancer properties of fruits and vegetables" HortScience 35: 573-575.

WFP (2007) World Food Programme Annual Report. Washington, UN.

WHO (2010) World Health Organization: "Diet, Nutrition and the Prevention of Chronic diseases.

WHO/FAO (2015) Diseases, Report of a Joint, expert consultation, WHO Technical Report Ser. 916 World Health Organization, Geneva.

Chapter 5

Disease Resistance Breeding in Tomato (*Solanum lycopersicum* L.)

Ramavath Ramesh Babu, G. Indraja, Shubham Singh and Raman Selvakumar

ABSTRACT

Tomato (Solanum lycopersicum L.) is an economically important solanaceous vegetable crop and also a research model plant for many biological and genetic studies. Commercially cultivated tomatoes are susceptible to many diseases which are the serious threats in production and productivity. Among the various biotic stresses, tomato leaf curl disease, groundnut bud necrosis virus, bacterial wilt, early blight, late blight and root knot nematodes has become major threat in production and causing considerable yield loss in the major tomato growing areas in the world. Traditional breeding efforts has been made to develop resistant breeding lines or cultivars to combat diseases. In past few decades, the development of "omics"science especially genomics and bioinformatics provided opportunity to breeders to integrate traditional breeding strategies with omics tools for resistance breeding. With the advent of DNA markers, markers assisted selection greatly facilitated the introgression of single or multiple resistance genes or QTL's in to commercially cultivated tomatoes for disease resistance. Marker assisted gene pyramiding facilitated increase in level of resistance and durability by combining two or more resistant genes. Marker assisted selection aided several advantages than conventional breeding such as increased efficiency and reliability of selection of resistance at seedling stage results in speed in breeding cycle. For disease resistance breeding various wild relatives of tomato have been utilized by introgressing resistance gene in to domesticated tomato. In this chapter, we summarize the currently available molecular markers that confer resistance against major tomato diseases, including late blight, early blight, Tomato Yellow Leaf Curl Virus (TYLCV), Tomato Spotted Wilt Virus (TSWV), Tomato Mosaic Virus (ToMV), verticillium wilt, fusarium wilt, leaf mold, root-knot nematodes, bacterial spot, and bacterial speck and also the use of resistance source(s), availability of resistance gene(s) conventional and marker assisted breeding approaches including relatively new tools for understanding and improving disease resistance in tomato.

Keywords: *Tomato, Disease resistance, Marker assisted selection, Breeding, Gene pyramiding, Resistance sources.*

Introduction

Tomato (*Solanum lycopersicum* L.) is an autogamous diploid (2n=2x=24) vegetable of the night shade family (Solanaceae) and the genus *Solanum*, which comprises of several species divided into two subgenera. The red-fruited species is *Eulycopersicon* and green fruited is *Eriopersicon*. Tomato is believed to be originated in mountainous regions of the Andes in Peru, Equador and Bolivia.

Tomato is one of the important vegetable crop cultivated in tropical and sub-tropical regions of the world. Globally, tomatoes are cultivated in an area of 4.85 million hectares with annual production of 182.30 million tonns (FAOSTAT, 2017). The major tomato growing countries are China, India, USA, Italy, Turkey and Egypt. In India, tomato ranks third in terms of area and production after potato and onion. India produces 22.33 million tonns of tomatoes from an area of around 0.8 million hectares (NHB data base 2017-18). Tomatoes are well adapted to tropical and subtropical regions of the world. Its versatility in fresh and processed forms has played a major role in its rapid and widespread adoption as an important nutritious food crops.

Tomatoes are consumed as fresh fruit, salad or in processed forms such as ketchup, juice, sauce, paste, soup, pickle and canned as whole fruit. Tomato is rich in vitamins A, C and provides sufficient amount of the antioxidant lycopene that helps to protect the body against cancer and heart diseases.

As a result, its cultivation has become more intensive and also highly remunerative during scarcity. Both acidic and regular segment varieties and F_1 hybrids are commercially adopted by the tomato growers in the country. Intensive and continuous cultivation of tomato in recent years has led to the prevalence of diseases and insect pests, which have become major production constraints in the country. Cultivated tomato has narrow genetic diversity because of extensive selection during breeding and domestication. Therefore, tomato crop is more vulnerable to disease epidemics.

The management of diseases by application of chemicals is very expensive and also ineffective in some pathogens. Therefore, it is important to exploit genetic resistance against these plant pathogens by developing cultivars/hybrids resistant to diseases. The resistant cultivars should also have other economic characters to make them viable for commercial exploitation. Success in disease-resistance breeding program depends on availability of resistant sources(s), which can be directly utilized after screening tomato germplasm under field and artificial inoculation conditions. Many efforts have been made for identification of sources of resistance and efficient utilization by many researchers using conventional and molecular breeding approaches. So far, conventional plant breeding approaches that utilize selection-recombination and selection cycles have led to improvement of tomato crop. However, these strategies are less precise and time-consuming, especially for improving complex quantitative traits. With the advent of molecular marker technologies, it is possible to localize genomic regions and assess their phenotypic effects on various quantitative traits.

Despite decades of breeding and selection efforts, there are still a number of fungal, viral, bacterial pathogenic diseases that challenge tomato production in various parts of the world. Recent developments in the area of plant genomics, including functional and structural genomics have paved ways for adoption of important tools for tomato breeding. These tools are being utilized for genetic analysis and crop improvement in a number of crop plants including vegetable crops.

Advent of DNA markers has helped in facilitating indirect selection of a phenotype based on the selection of marker linked to important traits thus helps in selection and speeding up selection cycles in crop improvement programs (Varshney and Tuberosa, 2007). They can be precisely and efficiently utilized for selection of traits with low heritability and gene introgression from native or exotic germplasm (Gupta *et al.*, 1996) and the identification of quantitative trait loci (QTLs) that control important agronomic traits (Dudley, 1993). Marker assisted selection provides plant breeders a quick and precise alternative strategy to conventional selection schemes which are based on extensive phenotyping for improving cultivars with respect to qualitative and quantitative traits like yield, adaptability and resistance to biotic and abiotic stresses.

In this chapter, we will discuss the use of resistant source(s), availability of resistant gene(s), conventional and marker assisted breeding approaches including new tools for understanding and improving disease resistance in tomato.

Breeding for Resistance to Fungal Pathogen

Tomato is affected by many 200 diseases. Among them major important fungal diseases which caus serious threats in tomato production are foliar diseases like late blight (caused by-*Phytophthora infestans*), early blight, (*Alternaria solani*), powdery mildew (*Oidium lycopersicum*), septoria leaf spot (*Septoria lycopersici*) and leaf mold (*Cladcsporium fulvum*). Other fungal diseases that infect tomato include fusarium wilt (*Fusarium oxysporium fsp. oxysporium*), verticilium (*Verticilium dahlia*), anthracnose (*C. coccodes* and *C. dematium*) and corky root (*Pyrenochaeta lycopersici*). These foliar diseases are serious threats which cause significant crop loss worldwide because of prevailing favourable weather conditions which determine the incidence and severity of the disease.

Late Blight (LB)

Phytophthora infestans is a plant destroyer best known for Irish famine in Ireland in 1845 which resulted in death and emigration of over one million people. Due to its devastating, economic and social impact, this disease has been the subject of intensified pathological and genetic research since the occurrence of the Irish potato famine in the 1840s. Late blight (*Phytophthora infestans* (Mont.) de Bary) is one of the most important production constraints of tomato in India and worldwide which causes substantial yield reduction up to 40 per cent -100 per cent (Nowakowska *et al.*, 2014; Ohlson and Foolad, 2015). The disease appears when weather conditions are cool and humid. *Phytophthora infestans* can destroy an entire unprotected tomato crop within 7–10 days in the field (Nowicki *et al.*, 2012).

The first complete dominant gene *Ph-1*, conferring resistance to *P. infestans* race T-0 was reported in *S. pimpinellifolium* accessions known as West Virginia 19 and 731 (Bonde and Murphy, 1952; Gallegly and Marvel, 1955; Peirce, 1971). In 1962, a late blight resistant cultivar Rockingham, carrying *Ph-1* gene was released (Rich *et al.*, 1962). *Ph-1* gene was mapped subsequently in this cultivar to the distal end of chromosome 7 using morphological markers (Peirce, 1971). Currently, T-0 is no longer considered as useful source of resistance for late blight in tomato as *Ph-1* has been overcome by new highly aggressive races of *P. infestans*.

A second tomato late blight resistance gene, *Ph-2*, was identified and mapped in chromosome 10 in *S. pimpinellifolium* accession known as West Virginia 700 (Gallegly and Marvel, 1955; Moreau *et al.*, 1998). This incompletely- dominant gene is also race specific and provides narrow resistance by slowing the rate of disease progress rather than blocking the disease. Further, *Ph-2* often fails in the presence of more aggressive *P. infestans* isolates (Goodwin *et al.*, 1995; Black *et al.*, 1996.). However, *Ph-2* has been incorporated into a number of cultivar such as Fline, Flora Dade, Heinz 1706, Campbell 28 and Europeel (Gallegly, 1960; Laterrot, 1994). Markers linked to LB resistance genes and resistance sources are listed in Table 5.1.

Table 5.1. List of Fungal Disease Resistant Genes and Linked Molecular Markers for Marker Assisted Selection

Resistant Gene (Chromosome)	*Marker*	*Marker Sequence*	*Marker Type*	*Reference*
a) Late blight (*Phytophthora infestans*)				
Ph-2	dTG63	F: CTACTCTTTCTATGCAATTTGAATTG R: AATAATTTTCAACCATAGAATGATT	CAPS (*Hinf1*)	Panthee *et al.* (2012)
		F: TCGATCGTATGTAGACGATG R: AGGCAAATCTTGAAGAAGCA	CAPS (*MspI*)	Jung *et al.* (2015)
	Dtg422	F-TGACATGAGAAGGAAAAGACTTAAG R-GTCAATAATTTTCAACCATAGAATGATT	CAPS (*Hinf1*)	Panthee *et al.* (2012)
Ph-3	Ph-3-GLR/S	F: GCTACCTTAATTATTTACAATTTTCCT R: TTGTGAAACAGAAAAGGTAAATATCA		Ren *et al.* (2019)
	NC-LB-9-6676	F: ACAGAAAAGTGCACGAAGTT R: ATTTGAATGTTCTGGATTGC	SCAR	Panthee *et al.* (2015)
b). Leaf mould (*Cladosporium fulvum*)				
Cf-9		F: TGGGAATTTACCCGAAAGAA R: CCCAATCATTCAGTGCGTTA	SCAR	Truong *et al.* (2011)
c) Verticillium wilt (*Verticillium* spp)				
Ve		F: CGAACTTGAC TACATTGACC CTG R: CAGTCTTGAAAGGTTGCTCAGCC	CAPS (*XbaI*)	Jung *et al.* (2015)
Ve		F: GAC TAC ATT GAC CCT GGG CTC TTG R: TGA GAG CAC CTT AAG CTT TTC AAT	CAPS (*XbaI*)	Kuklev *et al.* (2008)

Resistant Gene (Chromosome)	Marker	Marker Sequence	Marker Type	Reference
d) Fusarium wilt (*Fusarium spp*)				
I1	At2	F: CGAATCTGTATATTACATCCGTCGT R: GGTGAATACCGATCATAGTCGAG	SCAR	Arens *et al.* (2010)
I2	Z1063	F: ATTTGAAAGCGTGGTATTGC R: CTTAAACTCACCATTAAATC	SCAR	Arens *et al.* (2010)
I3	P7-43D	F: GGTAAAGAGATGCGATGATTATGTGGAG R: GTCTTTACCACAGGAACTTTATCACC	SCAR	Barillas *et al.* (2008)
I7	Solyc08g 077740	F: AAGAAGTTCCCTTCTTCCCTTA R: GGAATAACCAAGGGGGTGTT	CAPS (*Age*I)	Gonzalez-Cendales *et al.* (2015)
e) Grey leaf spot				
Sm	CT55	F- CATCTGGTGAGGCGGTGAAGTA R- TCCCCCAAACAAAACAGTAATA	CAPS (*Dde*I)	Ji and Scott (2009)
	Sm-InDel	F: CTACACTTTCTCGTTCCCAATG R: ATCGCCAAACCAATC AAATC	InDel	Su *et al.* (2019)
f) Bacterial Spot (*Xanthomonas campestris* pv. *vesicatoria*)				
Rx-3	Rx3-L1	F: CTCCGAGCGAAGAGTCTAGAGTC R: GAAGGCAAAAGGAAAAGGAGAAGGATGG	CAPS (*Bsr*BI)	Yang *et al.* (2005
Rx-4	pcc12	F: TCCACA TCAAA TGCGTTTCT R: TTCCAATCCTTTCCATTTCG	InDel	Pei *et al.* (2012)
Bs4	Bs4-A03/ B03	F: GGGTTGGAGTCCGAAGAGCAGG R: GACTAACCAACGCAAGTTATTGGACAGG	CAPS (*Msp*I)	Schornack *et al.* (2004)
g) Bacterial Speck (*Pseudomonas syringae* pv. *tomato*)				
Pto		F: ATCTACCCACAATGAGCATGAGCTC R: GTGCATACTCCAGTTTCCAC	CAPS (*Rsa*I)	Yang and Francis (2005)

A broad spectrum resistance gene from *S. pimpinellifolium* (L3708), *Ph-3* was identified and is most effective LB resistance gene among the known genetic sources, which confers incomplete resistance against a wide range of *P. infestans* isolates (Black *et al.*, 1996; Kim and Mutschler 2006; Zhang *et al.*, 2013).

Thus so far, no major LB resistance gene has been identified and characterized in other wild species of tomato. However, several QTL's conferring broad spectrum resistance have been identified in tomato wild species *S. habrochaites* (Robert *et al.*, 2001; Brouwer *et al.*, 2004), *S. pennellii* and *S. pimpinellifolium* (Frary *et al.*, 1998) Table 5.1a. Several breeding lines carrying *Ph-2, Ph-3* and *Ph-2+Ph-3* have been developed conferring resistance to late blight in tomato (Hanson *et al.*, 2016). Recently, ICAR-Indian Horticultural Research Institute, Bengaluru, India, has released a F_1 hybrid Arka Abhed resistant to late blight conferred by *Ph-2+Ph-3* gene. At ICAR-IIHR, several advanced breeding lines have been developed by pyramiding *Ty-2, Ty-3, Ph-2* and *Ph-3* genes for combined resistant to late blight and tomato leaf curl virus.

Table 5.1a. List of Important Tomato Disease and their Source of Resistance

Disease	*Resistant Sources*	*References*
Late blight (*Phytophthora infestans*)	*S. pimpinellifolium, S. cheesmaniae, S. pennellii*, L3708 Arka Abhed	Lukyanenko (1991); Smart *et al.* (2007)
Early blight (*Alternaria solani*)	*S. habrochaites* "LA1777", *S. pimpinellifolium, S. peruvianum*, NCEBR-1, NCEBR-2, Hybrids-Arka Rakshak, Arka Samrat, Plum Dandy, Mountain Magic, Mountain Merit and Mountain Supreme.	Sadashiva *et al.* (2017); Gardner (1988, 2000)
Bacterial spot	*S. lycopersicum* "H7981" and *S. pimpinellifolium* (accessions "PI126932" and "PI128216"), *S. pennellii* "LA716	Wang *et al.* (2012); Astua-Monge *et al.* (2000)
Bacterial Specks	*S. habrochites* "LA1777"	Thapa *et al.* (2015)
Leaf mold	S. pimpinellifolium, S. habrochaites	Parniske *et al.* (1997)
Fusarium wilt	*S. pimpinellifolium* accession "PI79532", *S. pennellii* "LA716", *S. pennellii* "PI414773" Pan American	Bohn and Tucker (1939); Scott and Jones (1989); Lim *et al.* (2006); Gonzalez-Cendales *et al.* (2015)
Verticillium wilt	*S. pimpinellifolium, S. esculentum "Peru"*	Kuklev *et al.* (2008)
Buck eye rot	*S. pimpinellifolium* "EC54725"	Rattan and Saini (1979)
Sclerotium wilt	AVTO1173	World Vegetable Centre, Taiwan
Anthracnose	PI 272636	Barksdale (1972)
Septoria leaf spot	S. *habrochaites*, S. *peruvianum*, S. *pimpinellifolium* and S. *glandulosum* "PI 126448"	Kurozawa and Balmer (1977); Kalloo (1986)
Gray leafspot (*Stemphylium solani* Weber)	S. *pimpinellifolium*, Florida Basket, Florida Lanai, Florida Petite and Flora-Dade	Andrus *et al.* (1942); Stevens and Rick (1986); Augustine *et al.* (1998); Hernandez (1985)
Corky root or brown root rot (*Pyrenochaeta lycopersici*)	S. *hirsutum* (*PI*129157), S. *peruvianum*, S. *chilense*, PI 262906, PI 198601, Mobohir 72-12 and Moboglan 2	Laterrot (1978); Veliev and Sviridova (1987); Jones *et al.* (1989)

Early Blight

Early blight is caused by *Alternaria solani* and *A. tomatophila* which can infect leaves, stems and fruits which eventually leads to fruit rot. Cultivated tomato is mostly susceptible to *Alternaria solani* which can cause up to 80 per cent reduction in yield (Yadav and Dabbas, 2012) and very few resistance accessions are available. Previous efforts identified resistance to early blight in PI 127827, an accession of *Solanum habrochaites* (Locke 1949); Targinni Red, Devon Surprise (Moore and Reynard 1945); Red cherry (Lodha,1977); *Solanum habrochaites* line PI 134417 (Datar and Mayee 1980); *Solanum habrochaites* PI 126445 (Gardner, 1984) and NCEBR-1 and NCEBR-2 (Gardner, 1988). A number of commercial cultivars with moderate but sufficient level of resistance have been reported and released (Table 5.1a).

S. habrochaites line PI 134417 was found to be highly resistant to both leaf blight and stem canker phases of *Alternaria solani* under severe epiphytotic conditions. *L. pimpinellifolium* showed moderate resistance to leaf-blight and was resistant to stem canker. Early blight resistant sources have been identified in wild relatives *viz.*, *S. habrochaites*, *S. pimpinellifolium* and *S. peruvianum*, and utilized to develop resistant breeding lines in tomato breeding program (Foolad *et al.*, 2008). *S. habrochaites* is the richest source of resistant genes among wild species of tomato, but these species are late maturing, indeterminate and low yielding. Some highly-resistant *S. habrochaites* accessions are: PI127827, PI126445, PI390513, PI390514, PI390516, PI390658, PI390660, PI390662, PI390663, LA2100. Amarnath *et al.* (2019) identified six genotypes (NCEBR-1, NCEBR-4, Arka Alok, Arka Rakshak, Arka Saurabh and *S. habrochaites* accession- LA1777) as highly resistant against early blight through detached leaf assay.

Several advanced breeding lines resistant to early blight have been developed using resistant wild species (Foolad *et al.*, 2000; Sadashiva *et al.*, 2017) (Table 5.1). Breeding for early blight resistance through conventional approach is difficult and challenging because of the polygenic (QTLs) nature of resistance and the linkage of resistance with several undesirable traits such as late flowering and fruit maturity. Chaerani and Voorrips (2006) attempted a cross between *S. habrochaites* and *S. lycopersicum* to obtain resistance, but it was observed that resistant plants were late in maturity and low in yield. In some cases where breeding lines were derived from *S. pimpinellifolium* accession LA2093 and tomato advanced breeding line NCEBR-1, resistance to early blight was high with desirable traits. However, some recombinant inbred lines (RILs) that were derived from the cross of mention species for cross and *S. lycopersicum* tomato breeding line were reported with EB resistance and other desirable characteristics including high yield, earliness in maturity and small plant size (Ashrafi and Foolad, 2015).

It has been reported that younger leaves are more resistant to early blight compared to older leaves because of presence of high total sugar content and glycoalkaloids (solanine, chaconine and solanidine) (Sinden *et al.*, 1972; Rotem, 1994). It is possible that a late maturing tomato line may appear early blight resistant without even having any resistant genes or QTLs because of the high sugar content and glycoalkaloids (Johanson and Thurston, 1990; Chaerani and Voorrips, 2006). A negative correlation between early blight resistance and early maturity made it difficult to breed early blight resistance using conventional methods.

Advancement in breeding for early blight resistance has been very slow due to lack of effective resistance genes in cultivated tomato (Vakalounakis 1983; Poysa and Tu 1996; Banerjee *et al.*, 1998; Vloutoglou 1999) and polygenic nature of inheritance of the resistance (Barksdale and Stoner 1977; Maiero *et al.*, 1989; Thirthamallappa and Lohithaswa 2000).

Resistance to early blight is mainly governed by polygenes and several QTL's have been mapped to different location on chromosomes with the help of DNA markers. Many efforts are being conducted to identify markers linked to QTL governing early blight disease resistance. Unfortunately, there were no success in developing molecular markers linked to QTL for the selection of early blight resistance because of minor QTL were utilized in breeding programs. The QTLs associated with EB resistance that have been identified so far are summarized.

Leaf Mold

Leaf mold, *Cladosporium fulvum* (*Fulvia fulvum*), is a serious threat of tomato cultivation especially of greenhouse tomatoes, which may cause yield loss up to 40–90 per cent. Several resistant cultivars/breeding lines have been developed in Canada (Vagabong, Vetomold, Vulcan, Vinequeen), Great Britain (Eurocross A, Eurocross B, Paghamcross, Antimold B, Pipo), Netherlands (Sonato, Sobeto, Revermun, Extase, Panase), USSR (Pionerskii, Leningradskii osennii, Dunai, Solnyshko, Turist 714) and in other countries (Table 5.1). But due to mutation, more aggressive pathogenic races of the fungus breakdown the resistance and cultivars become susceptible to new pathogenic races soon after release. Resistance to *C. fulvum* was governed by 24 dominant genes which are designated from *Cf-1* to *Cf-24* (Kerr *et al.*, 1971; Laterrot 1981). Gene *Cf-2* had shown immunity to two races of pathogen. Gene *Cf1* has broad spectrum resistance against most of the pathogenic races.

Gene *Cf-4* resistant to many races of pathogen including 6 and 7 is located on chromosome 1 (Kerr and Bailey 1964) and gene *Cf-2* located on chromosome 6 (Kerr *et al.*, 1980). The locations of *Cf-1* and *Cf-4* on chromosome 1 and *Cf-3* on chromosome 11 have been verified. *Cf- 5* and *Cf -21* are placed on chromosome 4, *Cf- 6* and *Cf- 11* on chromosome 12, *Cf-7* and a *Cf-8* on chromosome 9, *Cf-9* on chromosome 10, *Cf-10* on chromosome 7, *Cf-14* on chromosome 3, *Cf-19* and *Cf-20* on chromosome 2, *Cf-22* on chromosome 1 and *Cf-24* on chromosome 5 (Kanwar *et al.*, 1980a,b; Kalloo 1986). The resistance to many races of pathogen governed by the gene *Cf -1*, which is only partial, can be considered as race-nonspecific or horizontal resistance to *C. fulvum* (Boukema 1974).

Most of the sources of resistance have been found in wild species: Genes *Cf-2*, *Cf-3*, *Cf-5* to *Cf-10* were derived from *L. pimpinellifolium* (Kerr 1977; Boukema 1978, 1981; Laterrot 1981); gene *Cf-4* was derived from *L. hirsutum* f. glabratum and *L. peruvianum*. *Solanum pennellii* according to data presented by Stamova *et al.* (1985) is an additional source of the gene *Cf -4* as well as *Cf-2*. Some of its lines were found resistant to a mixture of races 1, 3, 4 and 5. This resistance is controlled by a single dominant gene either identical to that controlling resistance to the new race 2, 3, 4, 5, 9 in tomato cv Bulgaria No.1 (derived from *L. chilense*) or an allele at the same locus (Stamova *et al.*, 1984). Besides "Bulgaria No.1", Japanese lines Japan 74S and Japan 76S are possessing resistance to the race 3, 5, 9 (Laterrot 1986). Gene *Cf-5* which is providing resistance to most races of pathogen (1 to 12), and *Cf-4*, resistant to races from 1 to 9 (Kerr *et al.*, 1971), have the most practical interest in breeding for resistance to *C. fulvum*. Gene *Cf-1* controlling moderate resistance to many races (uniform resistance) has some interest too.

There have been no adequate studies regarding other genes conferring resistance to this pathogen. Since new races appear so rapidly and resistance governed by a single resistant gene is not durable. The best strategy for achieving durable and broad spectrum resistance is pyramiding of multiple genes in a single cultivar or genotype for leaf mold resistance.

Anthracnose

Anthracnose in tomato is caused by several species of *Colletotrichum*, but majorly by *C. coccodes*, *C. dematium* (Barksdale, 1972) and *C. truncatum* (Saini *et al.*, 2017; Villafana *et al.*, 2018). It is a serious disease in processing tomatoes, where ripe fruits are held on the plant for synchronous harvesting resulting in post-harvest fruit rot, which severely affects marketability of the fruits and processing quality. This is the most significant ripe fruit rot in USA and as well as in India. Initially, symptoms appear on ripe fruit as small, sunken, water-soaked circular lesions. The lesions increase in area, become more depressed, and the central portion darkens. The darkened rotten area constitutes of many small, dark, fungal structures from which masses of spores are released in moist weather. As the fungus spreads within the fruit, a semisoft decay occurs. Anthracnose lesions on a fruit are often coalesces and result in rotting of fruits. The disease is developed rapidly when temperature is 25° to 30°C, high relative humidity (>90 per cent) with intensive light. In tomatoes, high level of resistance to six anthracnose fungi was found in accession PI 272636 (Barksdale, 1972), which has been introgressed to advanced breeding lines of tomatoes (Stevenson *et al.*, 1978). Resistance to anthracnose is mainly governed by polygenes with incomplete dominance and epistatic interaction (Miller *et al.*, 1983).

Powdery Mildew

Powdery mildew disease is caused by *Oidium lycopersici* and *Leveillula taurica* (Le/v.) Arnaud (asexual state *Oidiopsis taurica* (Le/v.) Salmon). It has become globally important fungal disease since 1986, when it was first reported in the Netherlands (Paternotte in 1988). Later this foliar disease was quickly spread over all tomato cultivated areas in the world. Most of the modern tomato cultivars are susceptible to powdery mildew; however, resistance to *Oidium lycopersici* has been reported in many wild species (Ciccarese *et al.*, 1998; Lindhout *et al.*, 1994).

Resistant to *Leveillula taurica* has been found in accessions of *S. lycopersicum var. cerasiforme* and many wild species *viz.*, *S. chilense*, *S. habrochaites* and *S. peruvianum* (Palti 1988; Hernandes and Stamova 1990). The dominant resistance gene *Lv* has been mapped on chromosome 12 in *S. chilense* accession LA1969 which confers highly resistance against *L. taurica* by inducing hypersensitive response (Yordanov *et al.*, 1975; Stamova and Yordanov 1987; Chungwongse *et al.*, 1997). A recessive *ol-2* gene (Ciccarese *et al.*, 1998; Bai *et al.*, 2008) has been identified in *S. lycopersicum var. cerasiforme* and mapped on chromosome 4 which confers incomplete resistance to *L. taurica* (Zheng *et al.*, 2013). In addition to major genes several QTL's have been reported which confer high resistance to *Oidium lycopersici* (Table 5.1a).

Fusarium Wilt

Fusarium wilt caused by *Fusarium oxysporum f. sp. lycopersici Sacc.* (Fol) is a devastating disease causing 10–50 per cent yield loss in many tomato growing areas of the world and is serious in a warm humid weather and in sandy soils with acidic soil pH (5.5). Optimum temperature for fusarium infection is around 28°C. The disease symptoms are yellowing of the leaves, starting with the lower leaves to upward. Infected leaves later show downward curling, followed by browning and discolouration of vascular tissue and drying.

Three races of *Fusarium*, race 1, 2, and 3, have been identified (Bohn and Tucker 1939; Grattidge and O'Brien, 1982). Resistance against Fol has been identified in various wild species of tomato. First resistance *I-1* gene was identified on chromosome 11 in *S. pimpinellifolium* accession "PI79532" which confers resistance against Fol race 1 (Bohn and Tucker 1939). Second, *I-2* resistance gene was identified on chromosome 11 which confers resistance to Fol race 2, in *S. lycopersicum* x *S. pimpinellifolium* hybrid "PI126915", and this gene was reported to encode a coiled coil, nucleotide-binding (CC-NB)-LRR resistance protein (Stall and Walter, 1965). A third resistance gene *I-3* was identified and mapped in chromosome 7 in *S. pennellii* "LA716" which provides resistance to Fol race 3 (Scott and Jones, 1989) and this encodes a protein called S-receptor-like kinase (SRLK) (Lim *et al.*, 2008; Catanzariti *et al.*, 2015). Furthermore, a partial dominant resistance gene *I-7* was identified on chromosome 8 in *S. pennellii* "PI414773" which provides broad spectrum resistance to Fol race 1, 2, and 3, and encodes a typical extracellular LRR receptor-like protein (LRR-RLP) (Lim *et al.*, 2006; Gonzalez-Cendales *et al.*, 2015). Molecular markers for these resistance genes and source of resistance are listed in Tables 5.1 and 5.2, which can be used for marker assisted breeding for fusarium wilt resistance. Five tomato breeding lines (CLN3241S, CLN3241R, CLN3241P, CLN3241Q and CLN3241H-27) resistant to Fol race1 and race 2 were developed using conventional and marker assisted selection (Hanson *et al.*, 2016)

Verticillium Wilt

Tomato verticillium wilt, which is also known as vascular wilt disease, is a soil-born fungal disease caused by *Verticillium dahliae* and *V. alboatrum*, which are also known for the vascular wilt disease in over 200 dicotyledonous species (Fradin and Thomma, 2006).This disease can reduce the yield of susceptible genotype in field up to 40–47 per cent by race 1 isolates and by 19–31.2 per cent by race 2 isolates (Bender and Shoemaker 1984).

A single dominant gene *Ve* was identified and mapped on chromosome 9, and was reported to confer high resistance to *V. alboatrum* race 1 (Schaible *et al.*, 1951; Diwan *et al.*, 1999). The *Ve* locus was confirmed to contain two closely associated, inversely oriented genes, *Ve1* and *Ve2*, (Kawchuk *et al.*, 2001; Fradin *et al.*, 2009). *Ve1* resistance gene is reported to encode for extracellular LRR receptor-like protein class of disease resistance proteins (Kawchuk *et al.*, 2001), which is critical for switching on effector-triggered immunity in the host plant. Fradin *et al.* (2009) reported that *Ve1*-mediated resistance signalling partially overlaps with signalling mediated by Cf (*C. fulvum*) proteins, type members of the receptor-like protein class of resistance proteins. Jung *et al.* (2015) developed gene-based molecular markers (Table 5.2) for verticillium wilt resistance which can be used for marker assisted selection for disease resistance in tomato breeding programs.

Table 5.2. Molecular Markers Associated with Resistance to Root-knot Mematodes in Tomato

Gene	*Marker*	*Marker Type*	*Marker ID*	*Forward/Reverse*	*Restriction Enzyme*	*Product Size (bp)*		*Reference*
						R	*S*	
Mi	CAPS	Co-dominant	Aps	F- GGAACGTGGGTAGCATATGA R- GCCAATGCTCATCAATGTGA	Taq I	1 kb and 0.6 kb	Uncleaved (N-1.6 KB)	Ammiraju *et al.* (2003)
Mi	CAPS	Co-dominant	C8B	F- TACCCACGCCCCATCAATG R- TTGCAAGAGGGTGAATATTGAGTGC	Taq	115, 292	407	Ammiraju *et al.* (2003)
Mi	CAPS	Co-dominant	REX-1	F- TCGGAGCCTTGGTCTGAATT R- GCCAGAGATGATTCGTGAGA	Taq	N-750 570, 162	Ucleaved	Ammiraju *et al.* (2003)
Mi	CAPS	Co-dominant	CT119	F- TCAGGTATCGAACCAAAC R- TAAAAGGTTCATCCTAATA	Mse I	N-450 280, 170	250, 170	Ammiraju *et al.* (2003)
Mi-1	SCAR	Co-dominant	Mi23	F-TGG AAA AAT GTT GAATTT CTT TTG R-GCA TAC TAT ATG GCTTGT TTA CCC	380		430	Seah *et al.* (2007)
Mi-1	SCAR	Co-dominant	PMi	F GGTATGAGCATGCTTAATCAGAGCTCTC R-CCTACAAGAAATTATTGTGCGTGTGAATG	550		350	Arens *et al.* (2010)
Mi-3	SCAR	Co-dominant	TG-180	F-ATACTTCTTTGCAGGAACAGCTCAC R-CACATTAGTGATCATAAAGTACCAG	1.2 kb		0.9 kb	Yaghoobi *et al.* (2005)
Mi-3	SCAR	Co-dominant	TG-263	F-GCTGAGAAATAAAGCTCTTGAGG R-TACCCTTAATGCTTCGGCAGTGG	0.9 kb		0.75 kb	Yaghoobi *et al.* (2005)
Mi-HT	SSR	Co-dominant	SSR-W415	F: AAGTCTTATCTAATTGCCTAT R: ATTTCCGTAATGATATGATCT	670		550	Wang *et al.* (2013)
Mi-HT	CAPS	Co-dominant	Mi	F: GACCAACTTCCCCCTCCAAGATCA R: CTGTTCTAACCGGTTGCGGAGTAAA	Kpn I	80/570	650	SGN
Mi-HT	CAPS	Co-dominant	W737	F: GCCCTGATGAACTATTAGATGTC R: ACTCGCAACTTTACAAGATCC	Hind II	281/638/919	919	Wang *et al.* (2013)

Table 5.2a. RKN Resistant Sources in Tomato

Sl.No.	Varieties/F_1 Hybrids	Sources
1	Arka Vardhan	IIHR, Bengaluru
2	Pusa-120, Pusa Hybrid-2 and Pusa Hybrid-4	IARI, New Delhi
3	Punjab NR-7, Sel-1-6-1-4	PAU, Ludhiana
4	Hisar Lalit	HAU, Hisar
5	Narendra Tomato-2	NDUA and T, Faizabad, UP
6	CO-3	TNAU, Coimbatore, TN
7	TH-802, TH-2312	

Septoria Leaf Spot

Septoria leaf spot (*Septoria lycopersici* Speg.) is one of the major fungal disease of tomato throughout the world. The fungus infects the leaves, stems, and the green fruits. Cultivars "Targinnie Red" (Andrus and Reynard 1945) and "Bounty", "Campbell 1327", "Heinz 1350" and "Rutgers" (Benedict, 1972) were found to be resistant. Resistance was found in wild species (Table 5.2a): *S. habrochaites, S. peruvianum, S. pimpinellifolium and S. glandulosum* PI 126448 (Kurozawa and Balmer 1977). The resistance is governed by a single dominant gene "*Se*" (Barksdale and Stoner 1978). An advanced breeding line AVTO1173 was found to be highly resistant to septoria leaf spot (Gul *et al.*, 2016). Satelis *et al.* (2010) identified five lines with high degree of resistance in *S. peruvianum* accessions (PI-306811, CNPH-1036, LA-1910, LA-1984 and LA-2744). The resistance sources and to septoria leaf spot is listed in Table 5.2a.

Gray Leaf Spot

Gray leaf spot is important fungal disease caused by *Stemphylium solani* Weber which is epidemic in some areas where continuous high temperature and humidity occur. Tomato wild species *S. pimpinellifolium* showed high level of resistance to *S. solani* (Andrus *et al.*, 1942; Stevens and Rick 1986). Resistance to grey leaf spot is governed by a single dominant gene *Sm* which is closely linked to the *I* gene and mapped on short arm of chromosome 11 (Yang *et al.*, 2017). Marker linked to incomplete dominant gene *Sm* is listed in Table 5.1. Various cultivars with resistance to grey leaf spot disease have been developed: Florida Basket, Florida Lanai, Florida Petite (Augustine *et al.*, 1981). Cultivars Campbell 28, Rotella, Caraibe, Ont. 7620 and Flora-Dade were found to be highly resistant to all three races and were recommended as sources for resistance in breeding (Hernandez, 1985).

Breeding for Resistance to Root-Knot Nematodes

Root-Knot Nematode

Another major production constraint is root-knot nematodes (RKN) (*Meloidogyne spp.*) and extent of economic losses caused in tomato by RKN infestation range between 20.6-85 per cent (Sasser, 1989). Use of nematicides in commercial fields is impractical due to high costs and to health, soil and environment hazards. Breeding resistant cultivars is the only practical approach to manage these nematodes in fields. Exploitation of hybrid vigour for yield improvement in tomato began a century ago (Hedrick and Booth, 1907) and till recently many varieties/hybrids resistant to bacterial wilt and root knot nematodes have been released from private as well as public sector, but varieties/hybrids having combined resistance to both these diseases are not available. Nematodes account for an estimated 14 per cent of all worldwide plant losses, which translates into almost $100 billion dollars annually. So far, root-knot nematodes are the most common and destructive nematode pathogens. They produce some of the most dramatic symptoms and can substantially reduce crop yields. Root-knot nematodes are found in all agricultural regions worldwide. They can survive in temperate climates and can devastate crops grown in the tropics. Most root-knot nematodes also have extremely wide host ranges. Although it is difficult to ascertain the number of hosts for any one root-knot nematode species, it is likely that some root-knot nematodes can survive on hundreds of different plant species. This can make it extremely difficult to control a root-knot nematode problem, particularly if nematode can survive on weeds. In addition, root-knot nematodes have repeatedly been shown to predispose their host plants to infection by other crop pathogens, increasing the potential for crop loss.

Symptoms

Below Ground Symptoms (on plant Roots)

Root-knot nematode symptoms on plant roots are dramatic. As a result of nematode feeding, large galls or "knots" can form throughout the root system of infected plants. Severe infections result in reduced yields on numerous crops and can also affect consumer acceptance of many plants, including vegetables. The degree of root galling generally depends on three factors: nematode population density, *Meloidogyne* species, race host plant species and even cultivar.

Above Ground Symptoms

While the most diagnostic root-knot nematode damage occurs below ground, numerous symptoms can also be observed above ground. Because galled roots have only limited ability to absorb and transport water and nutrients to the rest of the plant, severely infected plants may wilt even in the presence of sufficient soil moisture, especially during the afternoon. Plants also may exhibit nutrient deficiency symptoms (chlorosis) because of their reduced ability to absorb and transport nutrients from the soil. Stunting is frequently observed on host crops grown in root-knot nematode-infested fields, and crop yields are reduced.

Resistance Sources

Root knot nematodes (RKN) resistance in tomato was first observed by Bailey (1941) in *S. peruvianum* (PI-128657). Using embryo rescue, Smith (1944) introduced this trait into *S. lycopersicum*, by crossing 'Michigan State Forcing' with 'PI 128657'. Three major resistance genes have been identified in the green-fruited tomato wild species *S. peruvianum*, including *Mi-1*, *Mi-3* and *Mi-5*, which have been mapped to tomato chromosomes 6, 12 and 12, respectively (Milligan *et al.*, 1998; Yaghoobi *et al.*,

1995; Veremis and Roberts, 1996). One heat stable gene *Mi-9* has been identified in *S. arcanum*, and mapped on chromosome No. 6 (Ammiraju *et al.*, 2003). Gene *Mi-1*, introgressed from *S. peruvianum* L., confers resistance to three species of root-knot nematodes, *Meloidogyne arenaria*, *M. incognita* and *M. javanica*, as well as to the potato aphid (*Macrosiphum euphorbiae*) (Dropkin, 1969; Rossi *et al.*, 1998).

Inheritance

Inheritance of heat stable resistance to the root knot nematode has been reported to be monogenic (single dominant gene) in *Lycopersicon peruvianum* (PI 270435 crossed with PI 126440) (Cap *et al.*, 1993). Shrestha *et al.* (2012) also reported single dominant resistance in Pusa 120.

Molecular Markers

Four CAPS markers (Aps, C8B, REX-1 and CT119) linked to root knot nematodes have been identified (Ammiraju *et al.*, 2003) and four SCAR markers (*Mi-23, PMi, TG-180* and *TG-263*) linked to root knot nematodes have also been identified (Seah *et al.*, 2007; Arens *et al.*, 2010; Yaghoobi *et al.*, 2005). One SSR marker (SSR-W415) linked to *Mi* gene also has been identified (Wang *et al.*, 2013).

Breeding for Bacterial Diseases

Bacterial Wilt

The major constraint to tomato production in India is bacterial wilt caused by *Ralstonia solanacearum*. This pathogen has a host range of more than 200 species covering over 50 families (Aliye *et al.*, 2008) and cause up to 75–100 per cent tomato yield loss in the lowland and highland tropics and subtropics (Denny, 2006). This disease is most severe in tropical and subtropical climates with high rainfall and warm temperatures (high temperatures and moist soil). The disease develops slowly when the soil temperature is lower than 20°C or soil moisture is low. The pathogen prefers acidic soils (soil pH < 7.0). The disease can occur on all types of soil, including sandy and clay types. The presence of root-knot nematode (*Meloidogyne spp.*) will accelerate disease development. High pathogen variability, high survival rate under diverse environmental conditions and its extremely wide host range renders the disease difficult to control. Chemical treatments are short lived and less effective. Therefore, control measures based on the use of resistant cultivars remain the most effective, economical and environment friendly method (Denny, 2006).

Symptoms

Wilting first appears on the younger leaves of plants during hot day time temperatures. The infected plants may recover, temporarily, in the evening, when temperatures are cooler. A few days later, a sudden and permanent wilt occurs. The roots and lower portion of the stem have a browning of their vascular system. The invaded roots may rot due to infection from secondary bacteria. Diseased stems that are cut and placed in a small container of water will show yellowish or greyish bacterial ooze coming from the cut end. Symptoms of this disease are distinguished from those of bacterial canker, which causes leaf chlorosis, stem cankers, and bird's eye spots on fruits. Bacterial wilt symptoms are distinguished from those of fusarium wilt because of the rapidity of the wilt, under favourable conditions, for the former, and the drier, firmer stem rot of the latter.

Resistance Sources

'Hawaii 7996' (*Solanum lycopersicum*) is one of the most stable sources of resistance against *R. solanacearum* bred by J.C. Gilbert in the 1970s (Wang *et al.*, 1998). AVRDC developed bacterial wilt resistance breeding lines (CLN1462, CLN1463) with heat-tolerance (Opena *et al.*, 1989). Numerous sources for resistance to bacterial wilt have been found in *S. pimpinellifolium*, *S. lycopersicum var. cerasiforme* and cultivated tomato since the first discovery of the resistant accession PI127805A (*S. pimpinellifolium*) in 1964 (Acosta *et al.*, 1964; Yin *et al.*, 2005; Yang and Francis, 2007; Alsam *et al.*, 2017). Most resistance in current resistant cultivars is derived from three major sources, PI127805A (*S. pimpinellifolium*), CRA66 (*S. lycopersicum var. cerasiforme*) and PI129080 (*S. pimpinellifolium*), and some other additional resistant sources (Hanson *et al.*, 1998). Inheritance: The resistance can be dominant, partially dominant or recessive depending on the resistant sources and bacterial strains used (Yue *et al.*, 1995; Li *et al.*, 2001; Wang *et al.*, 2004; Yang and Francis, 2007). The genetics of resistance are complex and controlled mostly by polygenes. Inheritance of bacterial wilt has been reported to be polygenic by Mangin *et al.* (1999) and Hanson *et al.* (1998) whereas single dominant resistance gene was reported by Scott *et al.* (1988) and Grimault *et al.* (1995).

Molecular Markers

At least seven QTL's on six chromosomes (3, 4, 6, 8, 10, and 12) from the resistant line Hawaii7996 (resistance derived from PI127805A) and three QTLs on three chromosomes (6, 7, and 10) from a *S. lycopersicum var. cerasiforme* accession L285 have been reported (Yang and Francis, 2007). The QTL's on chromosome 6 identified from both resistant sources are located in the same region, which could be a resistance gene cluster with different loci conferring resistance to different strains, or could be one locus conferring resistance to all three tested strains. This QTL shows a stable field performance of resistance (Wang *et al.*, 2004) and may be used for developing cultivars with resistance to bacterial wilt. Single nucleotide polymorphism (SNP) markers tightly linked to two QTLs *Bwr-6* (chromosome 6) and *Bwr-12* (chromosome 12) have been identified recently (Kim *et al.*, 2018). These markers will facilitate the process of developing resistant cultivars using MAS. Two major bacterial wilt resistance quantitative trait loci (QTLs), *Bwr-12* and *Bwr-6*, were identified in cv. 'Hawaii-7996' (Wang *et al.*, 2013). Two SCAR markers (TSCARAAT/CGA and TSCARAAG/CAT) linked to bacterial wilt resistance derived from the pair-cross between a BW resistant cultivar T51A and the susceptible cultivar T9230 were identified (Miao *et al.*, 2009). Another SCAR marker (SCU176-534) linked to bacterial wilt (*R. solanacearum*) resistance developed in line Hawaii-7996 using bulked-segregant analysis by cross between *S. lycopersicum* Hawaii 7996 (resistant parent) and *S. pimpinellifolium* WVa 700 (susceptible parent) (Truong *et al.*, 2015). Two SSR markers (SLM12-2 and SLM12-10) linked to bacterial wilt resistance have been identified (Wang *et al.*, 2013).

Table 5.3. BW Resistant Varieties/Lines in Tomato

Sl.No.	Varieties/F_1 Hybrids	Sources
1	Arka Alok, Arka Abha, Arka Abhijit, Arka Ananya, Arka Shreshta, Arka Rakshak (F_1), Arka Samrat (F_1)	ICAR-IIHR, Begaluru
2	Sakthi, Mukthi, Anagha, Vellayani Vijay	KAU, Kerala
3	Swarna Lalima, Swarna Naveen, Swarna Samridhi (F_1), Swarna Sampada (F_1), Swarna Deepti (F_1), Swarna Vijaya (F_1), Swarna Anmol (F_1),	HARP, Ranchi (Research station of ICAR-RCER, Patna)
4	Utkal Pallavi, Utkal Kumari, Utkal Urbashi	OUA and T, Bhubaneshwar
5	Megha, DMT-2	UAS, Dharwad
6	Sonali	BSKVV, Dapoli, MH
7	VC-48-1	AAU, Jorhat

Other Bacterial Disease of Tomato

Bacterial Speck

Bacterial speck caused by *Pseudomonas syringae* pv. *tomatoe* (Pst) is an economically important disease of tomato primarily in cool and moist conditions. Although this disease has been known since the early 1930s, it emerged as a devastating disease in late 1970s, when serious yield loss was recorded (Goode and Sasser, 1980). Disease symptoms include small (1 mm) black or brown necrotic lesions (specks) that become surrounded by chlorotic haloes, caused by the bacterial toxin coronatine (Boch *et al.*, 2002). Lesions also form on both unripe and ripe tomato fruit, and this manifestation of the disease can decrease marketability of the fruit (Jones, 1991). Incidence of the disease in the field varies from 15 per cent to 100 per cent (Zhao *et al.*, 2001; Wang *et al.*, 2015; Deng *et al.*, 2008). Sources with resistance to bacterial speck have been identified in several wild species and in cultivated tomato. Most studies suggest that resistance to bacterial speck is simply inherited. Four dominant or incompletely dominant genes have been reported so far. Genes *Pto* (also known as *Pto-1*) from *S. pimpinellifolium* accession PI370093, *Pto-2* from *S. pimpinellifolium* accession PI126430, and *Pto-3* from *S. habrochaites* accession PI134417 deliberate resistance to race 0, while gene *Pto-4* from PI134417 confers resistance to race 1 (Yang and Francis, 2007). Quantitative resistance has also been reported and QTL's conferring resistance to race 1 have been identified in *Solanum habrochaites* accession LA1777 (Thapa *et al.*, 2015). The resistance gene *Pto* has been cloned (Martin *et al.*, 1993) and can provide effective resistance against race 0 (Pedley and Martin, 2003). Therefore, it has been widely used in various breeding programmes and cultivars bearing this gene are commercially available worldwide (Yang and Francis, 2007). Gene based molecular markers for selecting the *Pto* gene have also been developed (Sun *et al.*, 2011) and have been used for MAS in breeding programs.

Bacterial Canker

Bacterial canker is a vascular disease of tomato caused by the gram positive bacterium *Clavibacter michiganensis* subsp. *michiganensis* (Cmm) (Davis *et al.*, 1984).

It was first conveyed in the USA in 1909 (Smith, 1910) and emerged as a devastating disease of tomato worldwide, which cause huge yield loss (Yang and Francis, 2007; Sen *et al.*, 2015; Chalupowicz *et al.*, 2017). Sources of resistance to bacterial canker have been identified in some wild species and incorporated into cultivated background in various countries including United States, Netherlands and Bulgaria (Yang and Francis, 2007; Sen *et al.*, 2015). *Solanum nigrum* was identified as a resistant source to bacterial canker and incorporated into cultivated background through backcrossing (Li *et al.*, 2012). Genetic analysis studies suggested that the resistance to bacterial canker is quantitatively inherited; therefore development of resistant cultivars can be facilitated by marker-assisted selection (MAS), which can help minimize the undesirable linkage drag from wild relatives and accelerate the breeding programme. Some molecular markers allied to the QTL's conferring resistance to bacterial canker have been identified in *S. peruvianum* accession LA2157 and *S. habrochaites* accession LA407 (Yang and Francis, 2007; Sen *et al.*, 2015). These markers might provide a tool for facilitating transfer of resistance from wild species into cultivated tomatoes. However, the linkage between the markers and resistance is not very tight. Therefore, fine mapping or cloning of genes/QTLs conferring resistance to the disease is necessary to identify molecular markers tightly linked to the resistance or develop functional markers of genes/QTLs.

Bacterial Spot

The causal agent of tomato bacterial spot is a complex of at least four species of Xanthomonas (*X. euvesicatoria*, *X. vesicatoria*, *X. perforans*, and *X. gardneri*) (Jones *et al.*, 2004) though recent studies suggest that *X. euvesicatoria* and *X. perforans* may be considered as a single species (Barak *et al.*, 2016; Cui, 2016). Bacterial spot is a dead pan disease of tomato especially in humid tropic region worldwide (Stall *et al.*, 2009; Jones *et al.*, 2014). The occurrence of the disease firstly noticed in China during mid-1940s (Li, 1991). The incidence of the disease varied from 29 per cent to 100 per cent (Sun *et al.*, 1991; Ding *et al.*, 1996; Li *et al.*, 1997; Wang and Ying, 2008; Guo *et al.*, 2008). Sources of resistance to the four species of bacterial spot have been identified and confirmed in both wild species and cultivated tomato germplasm (Yang and Francis, 2007; Scott *et al.*, 2013; Liabeuf *et al.*, 2015). Resistance in Hawaii-7998, Hawaii-7981, PI114490 (*S. lycopersicum var. cerasiforme*), and PI128216 (*S. pimpinellifolium*) has been incorporated into breeding programs and cultivars with resistance to at least one of the four species have been developed (Francis and Miller, 2005; Zhang *et al.*, 2009; Hutton *et al.*, 2010; Scott *et al.*, 2013; Liabeuf, 2016). Resistance to race T1 in Hawaii7998 is quantitatively inherited and a major QTL on chromosome 5 accounting for 41 per cent of phenotypic variation in an advanced backcross population has been identified (Yang *et al.*, 2005). A molecular marker linked to this QTL has been used for combining resistance to bacterial spot and bacterial speck (Yang and Francis, 2005) as well as MAS (Zhang *et al.*, 2009). Resistance to race T3 in PI128216 is conditioned by a single dominant gene Rx4 locating on a 45.1 kb region of chromosome 11 and a molecular marker tightly linked to this gene has been developed for MAS (Pei *et al.*, 2012). Resistance to race T3 in Hawaii7981 and LA1589 (*S. pimpinellifolium*) is also controlled by a single dominant gene mapped to the same region as Rx4 (Wang *et al.*, 2011; Sun *et al.*, 2011), which has been confirmed to be allelic to Rx4 in a later study (Zhao *et al.*, 2015). The RXopJ4 resistance locus from *S. pennellii* accession LA716 confers resistance to *X. perforans* (Sharlach *et al.*, 2013). Resistance to races T1-T4 and *X. gardneri* in PI114490 is quantitatively inherited and several QTL's have been identified (Scott *et al.*, 2003; Sun *et al.*, 2014). QTL's for resistance to *X. gardneri* in two *S. pimpinellifolium* accessions (LA2533 and LA1545) have also been identified (Liabeuf, 2016).

Breeding for Resistance to Tomato Viral Diseases

The tomato hosts more than 200 species of a varied variety of pests and pathogens that can cause significant economic loss. Among these, about 146 viruses belonging to 33 genera are reported to infect tomato around the world (Dhaliwal and Sharma, 2006). Some of the major ones are tomato leaf curl virus/tomato yellow leaf curl virus (ToLCD/TYLCD), groundnut bud necrosis tospovirus (GBNV), Tomato mottle virus (ToMoV), Tomato mosaic virus (ToMV), Cucumber mosaic virus (CMV), Pepino mosaic virus (PepMV), Tomato chlorosis virus (ToCV)/Tomato infectious chlorosis virus (TICV), Potato virus X, Potato virus Y, Capsicum chlorosis virus (CaCV), Watermelon bud necrosis tospovirus (WBNV), Tomato ring spot virus (TRSV) and Tomato torrado virus. In India two viral diseases *viz.*, Tomato leaf curl disease/Tomato yellow leaf curl disease and Groundnut bud necrosis disease are the major constraints in tomato production. These two viruses alone cause significant economic loss in tomato cultivation, which globally made them the top two viral diseases of tomato. In the coming paragraphs, these two diseases are described briefly.

Tomato Leaf Curl Disease/Tomato Yellow Leaf Curl Disease

Tomato (yellow) leaf curl disease is a very vicious disease of tomato caused by whitefly transmitted Begomoviruses (Family: Geminiviridae). They have twinned icosahedral (geminate) particles, and are either monopartite with a genome of a single circle of single-stranded DNA of about 2.7 kb, or bipartite with two circles of ssDNA both of about 2.7 kb. Begomoviruses are sometimes associated with satellite molecules of circular ssDNA; alphasatellites or/and betasatellites. The monopartite begomoviruses were originally considered to have originated in the Old World, while the bipartite begomoviruses were predominantly located in the New World. It was first noticed in 1930s in Israel and found to be caused by a whitefly transmitted virus in the1960s. Since then, Tomato leaf curl disease has been reported from across the world. It has been a global limit to tomato (*Solanum lycopersicum*) production since the 1980s (Moriones and Navas-Castillo, 2000). TYLCD/ToLCD can be characterized by the presence of yellowing, curling and cupping of leaves, severe stunting and abortion of flowers and fruits which leads to yield reduction of up to 100 per cent (Abhary *et al.*, 2007) on the infected susceptible tomato plants. To date, several hundred isolates of Tomato yellow leaf curl/Tomato leaf curl causing viruses have been sequence characterized from different parts of the world indicating high genetic diversity and wide geographical distribution of the viruses. Leaf curl disease on tomato in India was first reported in 1948 by Vasudeva and Samraj. The disease is now broadly spread in almost all tomato growing regions of the country and tomato production is brutally constrained due to its regular outbreaks. A total of ten monopartite tomato infecting Begomovirus species are known to be related with Tomato leaf curl disease in India. These include Tomato leaf curl Kerala virus (ToLCKeV), Tomato leaf curl Ranchi virus (ToLCRnV), Tomato leaf curl Patna virus (ToLCPaV), Tomato leaf curl Rajasthan virus (ToLCRaV), Tomato leaf curl Pune virus (ToLCPV), Tomato leaf curl Lucknow virus (ToLCLuV), Tomato leaf curl Bangalore virus (ToLCBaV), Tomato leaf curl Karnataka virus (ToLCKaV), Tomato leaf curl Joydebpur virus (ToLCJoV), and Tomato leaf curl Gujarat virus (ToLCGuV). The two predominant bipartite viruses infecting tomato are Tomato leaf curl New Delhi virus (ToLCNDV) and Tomato leaf curl Palampur virus (ToLCPalV).

Breeding for Resistance to Tomato Leaf Curl Disease

Chemical control of whitefly is practically impossible due to varied host range and fast mobility of the insect vector of Tomato leaf curl virus. Genetic resistance is the most auspicious approach to manage the disease and its deployment through breeding tomato cultivars resistant to the virus has increased the importance in recent years. Domesticated tomato is known to be susceptible to TYLCV infection, but resistance exists in wild tomato species (Ji *et al.*, 2007b). Sources of resistance to the virus were discovered in some wild tomato species including *S. pimpinellifolium*, *S. peruvianum*, *S. chilense*, *S. habrochaites* (syn. *L. hirsutum*), *S. cheesmaniae* and *S. pennellii*. Since early 1970s, classical breeding programmes have attempted to introduce Tomato leaf curl virus resistance from the related wild species to the commercial cultivars. Programs to develop Tomato leaf curl virus resistant/tolerant cultivars were started in Israel in 1974 and later in Egypt and France using several wild species. *S. pennellii* (LA 716) have shown multiple pest resistance owing to the presence of type IV glandular trichomes and the acyl sugars that they secrete. Banerjee and Kalloo (1987) screened a large number of germplasm and identified *S. hirsutum f glabratum* 'B6073', *S. hirsutum f typicum* 'A1904' and an accession of *S. peruvianum* as highly resistant to the virus.

Till date six independently inherited Ty resistance genes have been identified and mapped on to different chromosomes. So far, three TYLCV resistance genes have been cloned. *Ty-1* and *Ty-3* originated from *S. chilense* accession LA1969 and LA2779, respectively (Zamir *et al.*, 1994; Ji *et al.*, 2007a). They are two alleles of the same gene that is positioned on the long arm of tomato chromosome 6 and encrypts an RNA-dependent RNA polymerase (RDR) (Verlaan *et al.*, 2011, 2013). *Ty-2*, originating from *S. habrochaites* accession B6013 which is located on the long arm of chromosome 11 and encodes an NB-LRR gene (Yang *et al.*, 2014; Yamaguchi *et al.*, 2018). *ty-5* is a loss-of-function mutant allele of the pelota gene located on chromosome 4. The mutation in *ty-5* is caused by a T-to-G transversion in the coding region, which occured in cultivated tomato (Lapidot *et al.*, 2015). Some evidences suggested that *ty-5* is derivative of a complex of *S. peruvianum* accessions (Anbinder *et al.*, 2009). In addition to these cloned genes, two resistance loci *Ty-4* and *Ty-6* have been mapped. *Ty*-4 is recognized from *S. chilense* LA1932 which is located on the long arm of chromosome 3 and has a negligible effect toward TYLCV resistance (Ji *et al.*, 2009). Most recently identified *Ty-6* which shows resistance to TYLCV is located on the long arm of chromosome 10, found to be originated from *S. chilense* accessions LA1938 and LA2779 (Hutton and Scott, 2014).

Molecular markers linked to each of the Ty resistance gene have been developed (Table 5.1). Major focus in breeding programme is on the Introgressions of *Ty-1*, *Ty-2*, and *Ty-3* into cultivated tomato. Using marker assisted selection (MAS), the resistance genes have been successfully introgressed into the commercial cultivars. Punjab Agricultural University, Ludhiana has developed cultivars 'Punjab Varkha Bahar 1', 'Punjab Varkha Bahar 2' and 'Punjab Varkha Bahar 4' resistant to local isolates of Tomato leaf curl virus and MAS is being employed to develop the resistant cultivars with multiple T*y* genes.

TYLCV-related Tomato yellow leaf curl Sardinia virus replace resistance in *Ty-2* (Barbieri *et al.*, 2010) and its breakdown has been demonstrated recently by the isolate of the mild strain of TYLCV (TYLCV-Mld) (Ohnishi *et al.*, 2016). Under high disease pressure T*y*-1-mediated resistance is not suitable which leads to resistance breakage in some cases (García-Cano *et al.*, 2008). Resistance breakage facilitates TYLCD epidemics, which urges plant breeders to continuously search for effective novel sources of resistance in the wild tomato gene pool. Continuous effort from the breeders and pathologists led to the identification of wild tomato accessions exhibiting no TYLCD symptoms upon infection have been reported in a number of species, including *S. arcanum, S. cheesmaniae, S. chilense, S. chmielewskii, S. corneliomulleri, S. galapagense, S. habrochaites, S. neorickii, S. pennellii, S. peruvianum, and S. pimpinellifolium* (Ji *et al.*, 2007b; Vidavski *et al.*, 2008; De la Peña *et al.*, 2010; Pereira Carvalho *et al.*, 2010; Tomás *et al.*, 2011; Yan *et. al.*, 2018).

Bud Necrosis Disease of Tomato

Bud necrosis disease caused by groundnut bud necrosis tospoviruses (GBNV) (family Bunyaviridae; genus Tospovirus) are evolving as a substantial limiting factor for the sustainable production of tomato in India. The genome of tospoviruses is of negative/ambisense polarity and consists of three enveloped isometric RNA segments that are denoted as large (L, <"8.9 kb), medium (M, <"4.8 kb), and small (S, <"2.9 kb) according to their sizes. They are transmitted by thrips (Thysanoptera) in a propagative manner. Since the spotted wilt virus was originate in South Australia in 1915 and it has been found in all the other continents. Tospoviruses have arose as serious viral pathogens affecting the cultivation of several field and horticultural crops. It is one among the top ten most important viruses which adversely cause geater economic losses in tomato through out the world (Dhaliwal and Sharma, 2016). In the Indian subcontinent, due to their economic impact on a wide range of important crops, utmost of the research was carried out only in India. Historical perspective of disease symptoms similar to those induced by tospoviruses have been described in India since the 1960s on several vegetable crops such as brinjal, chilli, cowpea, pea, potato and tomato (Mandal *et al.*, 2012). Since Tospovirus was known as a monotypic genus until 1990, with Tomato spotted wilt virus as the sole species, tospoviral diseases described in India were initially considered to be caused by Tomato spotted wilt virus. It was in 1992 that based on serology, the bud necrosis disease of groundnut caused by a tospovirus different from Tomato spotted wilt virus (Reddy *et al.*, 1992). Groundnut bud necrosis virus was further confirmed as a distinct tospovirus based on nucleocapsid protein (N) gene sequence (Satyanarayana *et al.*, 1996). Subsequently, another distinct tospovirus, a new and unusual disease on watermelon characterized by leaf mottling and dieback of shoots was observed during 1991 and 1992 in parts of southern India (Singh and Reddy, 1996). A distinct tospovirus, Watermelon bud necrosis virus, was found to be associated with this disease (Jain *et al.*, 1998). To date, Groundnut bud necrosis virus, Peanut yellow spot virus, Watermelon bud necrosis virus, Iris yellow spot virus, Capsicum chlorosis virus (Mandal *et al.*, 2012) and Tomato spotted wilt virus (Renukadevi *et al.*, 2015) have been identified in the country.

Until the mid-2000, the occurrence of Groundnut bud necrosis virus was limited to a few regions. Beginning 2001, epidemics of Groundnut bud necrosis virus were reported in several tomato growing regions of India, including Hoskote and Nasik regions of Karnataka and Maharashtra States, respectively where up to 100 per cent disease incidence was recorded. Subsequently, GBNV is renowned as the most economically important tospovirus of tomato, as losses due to GBNV alone have been estimated at more than US$89 million per annum in Asia. Other Tospoviruses reported to infect tomato in India include Watermelon bud necrosis virus and Capsicum chlorosis virus (Kunkalikar *et al.*, 2011).

Resistance Breeding to Bud Necrosis Disease

On tomato, GBNV symptoms include necrotic rings on leaf; stem necrosis, concentric rings and patchy color on fruit (Manjunatha, 2008; Mandal *et al.*, 2012). Necrosis of the foliage often leads to collapse of a stem or of the whole plant resembling symptoms of blight. In general, tomato plants infected at an early stage often collapse and die (Akram *et al.*, 2012). Development of GBNV resistant cultivars has been identified as the most sustainable solution to reduce the virus incidence in tomato.

TSWV is one of the most widely spread tospovirus species worldwide and has been targeted for genetic resistance approaches. The search for natural resistance sources and resistance genes has intensified, and within the past three decades several resistance sources have been reported and incorporated into commercial tomato cultivars. Among these sources, recessive (*sw2*, *sw3*, and *sw4*) and dominant genes (*Sw1a* and *Sw1b*) have been reported; however, their resistance was quickly overcome upon challenge with TSWV. For the genes denoted *Sw-6* (UPV 32) and *Sw-7* (LA 1938), partial resistance to a narrow range of TSWV isolates has been observed. Because these reported resistance genes have not been characterized molecularly, it is not clear whether they represent different genes located on distinct chromosomes or represent different alleles from a well-known resistance gene cluster. The exemption is the *Sw-5* gene cluster, of which many homologs have been sequenced. This gene cluster originates from *Solanum peruvianum* and has been the most widely deployed resistance source against TSWV because of its durability and the ability to provide stable resistance against tospovirus species and isolates from different geographic locations. Although the *Sw-5* gene cluster was cloned and characterized in 2001, only recently have molecular markers that assist in breeding been reported. The *Sw-5* gene cluster is located on the telomeric region of the long arm of chromosome 9. Five paralogs, named *Sw-5a* to *Sw-5e*, have been reported, but only the *Sw-5b* copy has been proven to be functional and sufficient for a broad-spectrum resistance against tospoviruses. Interestingly and rather uniquely for plant viruses, *Sw-5b* confers resistance not only to TSWV but also to related tospovirus species such as TCSV, GRSV, Alstroemeria necrotic streak virus (ANSV), and Chrysanthemum stem necrosis virus (CSNV) and even to the less related Impatiens necrotic spot virus (INSV). All these tospovirus species are phylogenetically related and present within a putative American clade (Turina *et al.*, 2016).

The tomato cultivars when evaluated in field condition of India, the cultivars with *Sw-5* (BL 1022) and *Sw-7* (CK-12) genes were found highly susceptible to GBNV (Sain and Chadha, 2016). Search for resistant cultivars of GBNV has been carried out by several workers, EC 5888 was highly resistant while EC 8630 and EC 26512 were found to be resistant (Venkatramana *et al.*, 2011) and resistance in them was found to be governed by the single dominant gene. Sain and Chadha, 2016, detected a high degree of field resistance (>80 per cent) in seven lines of *S. peruvianum* (L00735, L00671, L00887, L06138), *S. chilense* (TL02226) and *S. pimpinellifolium* (L03708, TL02213). Recently, in ICAR-IIHR, Bengaluru, IIHR-2988 (*S. chilense*), IIHR-1940 (*S. peruvianum*) and IIHR – 2901 was identified as resistant while IIHR-2953, IIHR-2809 and IIHR-2101 as moderately resistant to GBNV after mechanical sap inoculation. To incorporate these resistances in different elite backgrounds of tomato effectively, more studies needs to be directed towards understanding the genetic basis of its inheritance. This will further help in determining the breeding strategies for the successful introgression of these genes in different backgrounds. These lines can also be utilized for the development of molecular markers linked to the GBNV resistance.

Tomato Mosaic Virus (ToMV) Resistance

ToMV with a positive sense RNA genome belongs to the genus Tobamovirus with the family Virgaviridae. They are often seen as a general mottling or mosaic appearance on foliage. When the plant is severely affected, leaves may look like to ferns with raised dark green regions. Leaves may also become stunted. Infected plants may have a severe reduction in fruit set and those that do set may be dotted with yellow blotches and necrotic spots while the interior of the fruit is brown. Stems, petioles, leaves and fruit may all show signs of infection. *Tm-1* and *Tm-2* two resistant genes which confer resistance againt ToMV have been introgressed in to the cultivated tomatoes. The *Tm-1* gene, which shows semi-dominant inheritance, was initially recognized from *S. habrochites* "PI126445" (Pelham 1966; Watanabe *et al.*, 1987). Tm-1 protein does not share any functional domain with formerly known resistance (R) proteins but physically binds to and functionally inhibits the replication proteins of ToMV (Ishibashi *et al.*, 2007). The *Tm-2* resistance gene found to confer high level of resistance compared to *Tm-1* and it was characterized in *S. peruvianum*. The *Tm-2* and *Tm-22* resistance genes are deliberated to be allelic (Pelham 1966; Young and Tanksley 1989), and the *Tm-22* resistance gene is more resilient than the *Tm-2* resistance gene (Fraser 1990). Hence in tomato breeding programms *Tm-22* is considered to be both practically and economically important and thus is engaged as a ToMV resistance source. Both *Tm-2* and *Tm-22* encode a member of the coiled-coil/nucleotide binding-ARC/LRR protein class of plant resistance (R) genes (Lanfermeijer *et al.*, 2003). The Tm-22 and Tm-2 open reading frames only differ by seven nucleotides, resulting in four amino acid differences at the protein level. Two of these differences are located in the nucleotide-binding site, and two are located in the LRR domain (Lanfermeijer *et al.*, 2005). ToMV allele-specific DNA markers derived from these above-mentioned gene sequences are listed in Table 5.4.

Table 5.4. Molecular Markers for Selecting ToMV Resistance Genes, *Tm1* and *Tm22*

Resistant Locus	*Marker Information*		*References*
Tm1	CCACTGTATGATTTCTGCTAGTGAA AGCTT TAACAAATATAAGAATAAAGAC GCAAGCTAAGGT TTACATATATGCC	SCAR (gene-based)	Ishibashi *et al.*, 2007
Tm2	F:GAGTTCTTCCGTTCAAATCCTAAGCTTGAGAAG R: CTACTACACTCACGTTGCTGTGATGCAC	CAPS (gene-based)	Lanfermeijer *et al.*, 2003 Lanfermeijer *et al.*, 2005
	SCN131000 (AccI) F: AGCGTCACTCCATACTTGGAATAA R: AGCGTCACTCAAAATGTACCCAAA	CAPS (linked marker)	Sobir *et al.*, 2000

Table 5.5. Molecular Markers for Selecting Resistant Tomatoes against TYLCV/ToLCV

Resistant Locus (Chromosome)	*Marker Information*		*References*
Ty1/3 (ch 6)	F: ATGAAGACAAAAACTGCTTC R: TCAGGGTTTCACTTCTATGAAT (SspI, BglII, or TaqI)	CAPS (gene-based)	Jung *et al.*, 2015
Ty3 (ch 6)	SCAR 1 F: GCTCAGCATCACCTGAGACA R: TGCAGGAACAGAATGATAGAAAA	SCAR (gene-based)	Dong *et al.*, 2016
Ty2 (ch 11)	T0302 F: TGG CTC ATC CTG AAG CTG ATA GCG C R: AGT GTA CAT CCT TGC CAT TGA CT	SCAR	Yang *et al.*, 2014
Ty4 (ch 3)	C2_At4g17300 (AflI) F: ATTTAACCGTGTCTGGGCAACTCAATGG R: GCTCACTTTGCAAATCACATCCCCATTTCACC C2_At5g60160 (NlaIII) F: TTCTCGCGGCCTTTTCTCCTC R: TCGTGATCGCAAACATATACTCGC	CAPS	Ji *et al.*, 2009

Resistant Locus (Chromosome)	Marker Information		References
ty5(ch 4)	F: TTCAAGTCCTTCTTCAAC R: ATAGATTTAAACAACAATTATAGAGATAAAA AAGTTACCTGT (Rsal)	dCAPS (gene-based)	Lapidot *et al.*, 2015
	F-GGTGCTCATGGATAGCTTAC R-CTATATAGGCGATAGCACCAC	(SSR) TM273	Kardivel *et al.*, 2013
Ty6 (ch 10)	UF_10.61192 F: CATAAAGTTCCGGCGAGTGT R: TCCATTCCAAACCAAGTGAAG (BssHII)	CAPS	Gill *et al.*, 2019

Table 5.6. Resistant Sources Identified for some other Viruses Infecting Tomatoes

Sl.No.	Viruses	Reported Resistant Source	References
1.	Pepino mosaic virus (PepMV)	*S. habrochaites* (LA 1731)	Ling and Scott, 2007
2.	Potato virus Y	*S. habrochaites* (L.03683 and L.03684), PI 247087	Oliveira *et al.*, 2017: Thomas and McGrath,1998
3.	Tomato mottle virus (ToMoV)	*S. chilense* (LA 1932, LA 1938, LA 1969, and LA 2779)	Griffiths and Scott, 1998
4.	Tomato chlorosis virus (ToCV)	*S. habrochaites* PI 127827,*S. lycopersicum* LT05, *S.peruvianum* (sensu lato) CGO 6711, *S. chilense* LA1967 and *S.*	Arcos *et al.*, 2018

Breeding Methods in Tomato

Utmost important breeding techniques used in order to improve tomato has been hybridization followed by pedigree selection. In various breeding methods the back cross method of breeding has been used to transfer the desirable traits from the wild species to cultivated varieties. Breeding methods like pedigree methods, single seed descent method in combination with the molecular approaches have been found to be a useful.

Introduction

- ✰ Sioux, HS 110, Roma, Pant T3, Arka Vikas, Arka Saurabh, S12, Pant Bahar, Narendra Tomato 2, Narendra Tomato 1 and Marglobe

Back Cross Method

- ✰ Efficaciously used in breeding varieties resistant to diseases such as *Fusarium, Verticilium, Stemphylium* and Root Knot Nematode
- ✰ Transgression of the desired genes from the wild related *Lycopersicon* species into the cultivated species (*S. lycopersicum*)

- ☆ Transfer leaf curl virus resistance from *Solanum hirsutum* f. *glabratum* to cultivated tomato

Pedigree Method

- ☆ Common breeding method in tomato
- ☆ Single plant selection is initiated in F_2
- ☆ In early generations, selection – highly heritable characters like growth habit, resistance to diseases, fruit characteristics, *etc.*
- ☆ Later generation- Low heritability like fruit yield and quality

Single Seed Descent Method

- ☆ Conceptualized by Brim (1966)
- ☆ Modified bulk method
- ☆ Now encouraged to be used by tomato breeder as generation can be advanced in the off-season
- ☆ Allows the maintenance of broad genetic base in advanced generation
- ☆ Mostly used for bacterial wilt resistant varieties

Interspecific Hybridization Method

- ☆ It has been used extensively in tomato due its narrow genetic basis
- ☆ Embryo rescue- to produce interspecific hybrids due to sexual incompatibility
- ☆ Backcross has also been used for the development of substitution lines, advanced backcross and backcross inbred lines
- ☆ These populations signify advantages for the mapping and introduction of QTL- exotic germplasm

Heterosis Breeding

- ☆ Heterosis in tomato is a product of mid 20th century
- ☆ First reported in 1907 (Hedrick and Booth, 1907)
- ☆ First step towards a large scale extension of hybrid tomato cultivars took place in Bulgaria in 1932
- ☆ First tomato hybrid cultivar "Karnataka" introduced by Indo-American Hybrid Seed Company (IAHS) in 1973.

Transgenics

- ☆ The first transgenic cultivar, 'Flavr-Savr'™ in 1994
- ☆ The development of transgenic plants has continued mainly with the introgression of genes-disease resistance and to fruit quality
- ☆ Flavr Savr (Longer Lasting Tomato)

- In tomato, enzymes poly galacturonase degrades pectin which is the major component of fruit cell wall
- This leads to the softening of fruit and a deterioration in fruit quality
- Transgenic tomatoes have been produced, which contain antisense construct of the gene encoding PG
- These transgenics shows a drastically reduced expression of PG and markedly slower fruit softening
- These tomatoes have about 2 weeks longer shelf-life than normal tomatoes
- Such tomatoes were approved for marketing in U.S.A. under the name "flavr savr slow fruit softening tomato" (1994)

Molecular Approaches

- Marker assisted selection
- Marker assisted backcrossing
- Marker assisted gene pyramiding
- Marker assisted gene introgression- ABQTL and Introgression library
- Early generation marker assisted selection
- Gene tagging/marker trait associations

Conclusion

Tomato cultivation has become intensive to meet the increasing demand for tomatoes. Intensive cultivation has resulted in various biotic stresses caused by virus, bacteria, fungal pathogen and root knot nematodes. Adoption of disease resistant genotypes is the most practical way to combat these biotic stresses. There is an immediate need to address new emerging diseases in order to maintain the productivity

REFERENCES

Abhary M, Patil BL, Fauquet CM (2007) "Molecular biodiversity, taxonomy, and nomenclature of Tomato Yellow Leaf Curl-like Viruses," in Tomato Yellow Leaf Curl Virus Disease: Management, Molecular Biology, Breeding for Resistance, ed. H. Czosnek (Dordrecht: Springer). 85–118.

Acosta JC, Gilbert, JC, Quinon, VL (1964) Heritability of bacterial wilt resistance in tomato. Proc Am Soc Hortic Sci 84: 455–462.

Akram M (2012) Biological characterization and variability of the nucleocapsid protein gene of Groundnut bud necrosis virus isolates infecting pea from India. Phytopathologia Mediterranea 266-275.

Aliye N, Fininsa C Hiskias Y (2008) Evaluation of rhizosphere bacterial antagonists for their potential to bioprotect potato (*Solanum tuberosum*) against bacterial wilt (*Ralstonia solanacearum*). Biological Control 47: 282-288.

Amarnath M, Ravishankar KV, Satish D, Sriram S, Sadashiva A T (2019) Screening of tomato genotypes against early blight pathogen *Alternaria solani* using detached leaf assay method. Pest Management in Horticultural Ecosystems 25 (1): 93-101.

Ammiraju JS, Veremis JC, Huang X, Roberts PA, Kaloshian I (2003) The heat-stable root-knot nematode resistance gene *Mi-9* from *Lycopersicon peruvianum* is localized on the short arm of chromosome 6. Theor Appl Genet 106: 478-84.

Anbinder I, Reuveni M, Azari R, Paran I, Nahon S, Shlomo H (2009) Molecular dissection of tomato leaf curl virus resistance in tomato line TY172 derived from *Solanum peruvianum*. Theor. Appl. Genet 119: 519–530.

Andrus CF, Reynard CB (1945) Resistance to Septoria leaf spot and its inheritance in tomatoes. Phytopathology 35: 16-24.

Andrus CF, Reynard GB, Wade BL (1942) Relative resistance of tomato varieties, selections and crosses to defoliation by *Alternaria* and *Stemphylillm*. USDA Circ 652: 23.

Arens P, Mansilla C, Deinum D, Cavellini L, Moretti A, Rolland S, van der Schoot H, Calvache D, Ponz F, Collonnier C, Mathis R, Smilde D, Caranta C, Vosman B (2010) Development and evaluation of robust molecular markers linked to disease resistance in tomato for distinctness, uniformity and stability testing. Theor. Appl. Genet 120: 655-664.

Arens P, Mansilla C, Deinum D, Cavellini L, Moretti A, Rolland S, Vander-Schoot H, Calvache D, Ponz F, Collonnier C (2010) Development and evaluation of robust molecular markers linked to disease resistance in tomato for distinctness, uniformity and stability testing. Theor. Appl.Genet 120: 655-664.

Ashrafi H, Foolad MR (2015) Characterization of early blight resistance in a recombinant inbred line population of tomato: II. Identification of QTLs and their co-localization with candidate resistance genes. Adv Stud Biol 7: 149-168.

Aslam MN, Mukhtar T, Hussain MA, Raheel M (2017) Assessment of resistance to bacterial wilt incited by *Ralstonia solanacearum* in tomato germplasm. J Plant Dis Prot 124: 585–590.

Astua-Monge G, Minsavage GV, Stall RE, Vallejos CE, Davis MJ, Jones JB (2000) Xv4-vrxv4: A new gene-for-gene interaction identified between *Xanthomonas campestris pv. vesicatoria* race T3 and the wild tomato relative *Lycopersicon pennellii*. Molecular Plant-Microbe Interactions 13(12): 1346-1355.

Augustine JJ, Harbaugh BK, Crill J (1981) Florida Basket. A dwarf tomato for hanging baskets. Circ Agric Exp Stn Univ Fla NS 283: 2.

Bai Y, Pavan S, Zheng Z, Zappel NF, Reinstädler A, Lotti C, De Giovanni C, Ricciardi L, Lindhout P, Visser R Theres K (2008) Naturally occurring broad-spectrum powdery mildew resistance in a Central American tomato accession is caused by loss of Mlo function. Molecular Plant-Microbe Interactions 21(1): 30-39.

Bailey D M (1941) The seedling method for root-knot nematodes resistance. *Proc Am Hort Sci.* 38: 573-575.

Banerjee MK, Chhabra ML, Saini PS (1998) Responses of tomato cultivars to Alternaria blight. Tests Agrochem Cultiv. 19: 50–51

Banerjee MK, Kalloo G (1987) Sources and inheritance of resistance to leaf curl virus in Lycopersicon. Theor. Appl. Genet 73: 707710.

Barak JD, Vancheva T, Lefeuvre P, Jones JB,Timilsina S, Minsavage GV, Vallad GE, Koebnik R (2016) Whole-genome sequences of *Xanthomonas euvesicatoria* strains clarify taxonomy and reveal a stepwise erosion of type 3 effectors. Front Plant Sci 7: 1805.

Barbieri M, Acciarri N, Sabatini E, Sardo L, Accotto GP, and Pecchioni N (2010). Introgression of resistance to two Mediterranean virus species causing tomato yellow leaf curl into a valuable traditional tomato variety. J. Plant Pathol 92: 485–493.

Barillas ACL, Mejia A, Sanchez-Perez DP, Maxwwell (2008) CAPS and SCAR markers for detection of I-3 gene introgression for resistance to *Fusarium oxysporum* f. sp. *lycopersici* race 3. *Rep.* Tomato Genet. Coop 11-17.

Barksdale TH (1972) Resistance in tomato to six anthracnose fungi. Phytopathology 62: 660-663.

Barksdale TH, Stoner AK (1977) A study of the inheritance of tomato early blight resistance. Plant Dis Rep 61: 63–65.

Barksdale TH, Stoner AK (1978) Resistance in tomato to *Septoria lycopersici.* Plant Disease Rep 62(10): 844-847.

Bender CG, Shoemaker PB (1984) Prevalence of *Verticillillm* wilt of tomato and virulence of *Verticillillm dahliae* race 1 and race 2 isolates in Western North Carolina. Plant Disease 68: 305-309.

Benedict WG (1972) Influence of light on peroxidase activity associated with resistance of tomato cultivars to *Septoria lycopersici.* Can J Bot 50: 1931-1936.

Black L L, Wang TC, Hanson PM, Chen J T (1996) Late blight resistance in four wild tomato accessions: effectiveness in diverse locations and inheritance of resistance. Phytopathol 86: S24.

Boch J, Joardar V, Gao L, Robertson TL, Lim M, Kunkel BN (2002) Identification of *Pseudomonas syringae* pv. *tomato* genes induced during infection of Arabidopsis thaliana. Molecular microbiology 44: 73-88.

Bohn GW, Thcker CM (1940) Studies on Fusarium wilt of tomato. I. Immunity in *Lycopersicon pimpinellifolium* Mill. and its inheritance in hybrids. Mo Agric Exp Stn Res Bull 311: 1-82.

Bonde R, Murphy E F (1952) Resistance of certain tomato varieties and crosses to late blight. Maine Agric. Exp. Sta. Bull 497: 5–15.

Boukema IW (1978) Races of *Cladosporium fulvum* (*Fulvia fulva*) in the Netherlands and resistant genes in the genus *Lycopersicon.* In: Genotype and environment in glasshouse tomato breeding. Proc Eucarpia Tomato Working Group, Leningrad. 150-155.

Brouwer DJ, Clair St (2004) Fine mapping of three quantitative trait loci for late blight resistance in tomato using near isogenic lines (NILs) and sub-NILs. Theor Appl Genet 108. 628-638.

Brouwer DJ, Jones ES, Clair St (2004) QTL analysis of quantitative resistance to *Phytophthora infestans* (late blight) in tomato and comparisons with potato. Genome 47: 475-492.

Cap GB, Roberts PA, Thomason IJ (1993) Inheritance of heat-stable resistance to *Meloidogyne incognita* in *Lycopersicon peruvianum* and its relationship to the Mi gene. Theor Appl Genet 85: 777–783.

Catanzariti AM, Lim GT, Jones DA (2015) The tomato *I-3* gene: a novel gene for resistance to Fusarium wilt disease. New Phytologist 207(1): 106-118.

Chaerani R, Voorrips R (2006) Tomato early blight (*Alternaria solani*): The pathogen, genetics, and breeding for resistance. J Gen Plant Pathol *72*: 335–347.

Chalupowicz L, Barash I, Reuven M, Dror O, Sharabani G,Garte- mann KH, Eichenlaub R, Sessa G, Manulis-Sasson S (2017) Differential contribution of *Clavibacter michiganensis ssp. michiganensis* virulence factors to systemic and local infection in tomato. Mol Plant Pathol 18: 336–346.

Chen CH, Sheu ZM, Wang TC (2008) Host specificity and tomato-related race composition of *Phytophthora infestans* isolates in Taiwan during 2004 and 2005. Plant Dis 92: 751-755.

Chunwongse J, Chunwongse C, Black L, Hanson P (1998) Mapping of *Ph-3* gene for late blight from *L. pimpinellifolium* L3708. Rep Tomato Genet Coop 48: 13-14.

Chunwongse J, Chunwongse C, Black L, Hanson P (2002) Molecular mapping of *Ph-3* gene for late blight resistance in tomato. J Hortic Sci Biotechnol 77: 281-286.

Chunwongse J, Doganlar S, Crossman C, Jiang J, Tanksley SD (1997) High-resolution genetic map of the *Lv* resistance locus in tomato. Theoretical and applied genetics *95*(1-2): 220-223.

Ciccarese F, Amenduni M, Schiavone D, Cirulli M (1998) Occurrence and inheritance of resistance to powdery mildew (*Oidium lycopersici*) in *Lycopersicon* species. Plant Pathol 47: 417-419.

Cui Z (2016) Detection and analysis of *Xanthomonas* associated with bacterial spot in tomato and pepper seeds [M.D. Dissertation]. China Agricultural University, Beijing, China.

Datar VV, Mayee CD (1980) Breeding for early blight resistance in tomato. Indian Phytopatho l33: 151.

Davis MJ, Gillaspie AG,Vidaner AK, Harris RW (1984) *Clavibacter* : a new genus containing some phytopathogenic coryneform bacteria, including *Clavibacter xyli* subsp. *xyli* sp. *nov.*, subsp. *nov.* and *Clavibacter xyli* subsp. *cynodontis* subsp. *nov.*, pathogens that cause ratoon stunting disease of sugarcane and bermudagrass stunting disease. Int J Syst Bacteriol 34: 107–117.

De la Peña R, Kadirvel P, Venkatesan S, Kenyon L, Hughes J (2010) "Integrated approaches to manage Tomato Yellow Leaf Curl Viruses," in Biocatalysis and Biomolecular Engineering, eds C. T. Hou and J.-F. Shaw (Hoboken NJ: John Wiley and Sons). 105–132.

Deng G, Qu X, Chen X R, Yang CD, Xue L (2008) Pathogen identification of bacterial leaf speck of tomato based on 16S rDNA, physiology and biochemistry in Gansu Province. Plant Prot 34: 47–51.

Denny TP (2006) Plant pathogenic *Ralstonia* species. In Plant-Associated Bacteria, ed. SS Gnanamanickam. Dordrecht, The Netherlands. 573-644.

Dhaliwal, Sharma (2106) "Breeding for Resistance to Virus Diseases in Vegetable Crops", *in* Innovations in Horticultural Sciences, *ed.* K V Peter (New India Publishing Agency, New Delhi, India). 303327.

Ding AY, Shi CK, Zheng JF, Zhu HC, Jin XD (1996) Preliminary study on bacterial spot to tomato in Shandong. *J Shandong Agric Univ*. 27: 207–208.

Diwan N, Fluhr R, Eshed Y, Zamir D, Tanksley SD (1999) Mapping of *Ve* in tomato: a gene conferring resistance to the broadspectrum pathogen, *Verticillium dahliae* race 1. Theor. Appl. Genet 98: 315-319.

Dropkin VH (1969) Cellular responses of plants to nematode infections. Annu Rev Phytopathol 7: 101-122.

Dudley JW (1993) Molecular markers in plant improvement: manipulation of genes affecting quantitative traits. Crop Sci 33: 660–668.

Fallik E, Bashan Y, Okon Y, Kedar N (1984) Genetics of resistance to bacterial speck of tomato caused by *Pseudomonas syringae* pv. tomato. Annals of applied biology 104(2): 321-325.

Foolad MR, Merk HL, Ashrafi H (2008) Genetics, genomics and breeding of late blight and early blight resistance in tomato. Critical Reviews in Plant Sciences 27(2): 75-107.

Foolad MR, Ntahimpera N, Christ BJ, Lin GY (2000) Comparison of field, greenhouse, and detached leaflet evaluations of tomato germplasm for early blight resistance. Plant Dis 84: 967–972.

Fradin EF, Thomma BPHJ (2006) Physiology and molecular aspects of Verticillium wilt diseases caused by *V. dahliae* and *V. alboatrum*. Mol. Plant Pathol 7: 71-86.

Fradin EF, Zhang Z, Juarez, Ayala JC, Castroverde CD, Nazar RN, Robb J, Liu CM,Thomma BP (2009) Genetic dissection of *Verticillium* wilt resistance mediated by tomato *Ve1*. Plant Physiol 150: 320-332.

Frary A, Graham E, Jacobs J, Chetelat RT, Tanksley SD (1998) Identification of QTL for late blight resistance from *L. pimpinellifolium* L3708. Tomato Genet Coop Rprt 48: 19–21.

Fraser RSS (1990) The genetics of resistance to plant viruses. Annu Rev Phytopathol *28:* 179-200.

Gallegly ME (1960) Resistance to the late blight fungus in tomato. Proc Campbell Soup Co., USA: 113–135.

Gallegly ME, Marvel ME (1955) Inheritance of resistance to tomato race-O of *Phytophthora infestans*. Phytopathol 45: 103–109.

García-Cano E, Resende RO, Boiteux LS, Giordano LB, FernándezMuñoz R, Moriones E (2008) Phenotypic expression, stability, and inheritance of a recessive resistance to monopartite begomoviruses associated with Tomato yellow leaf curl disease in tomato. Phytopathology 98: 618–627.

Gardner RG (1984) Use of *Lycopersicon hirsutum* PI 126445 in breeding early blight resistant tomatoes (Abstr). Hort Science 19(2): 208.

Gardner RG (1988) NC EBR-l and NC EBR-2 early blight resistant tomato breeding lines. Hort-Science 23: 779-781.

Gardner RG, Panthee D R (2010b) NC 1 CELBR and NC 2 CELBR: early blight and late blight resistant fresh market tomato breeding lines. Hort-Science 45: 975—976.

Gardner RG, Panthee DR (2010a) 'Plum Regal' fresh-market plum tomato hybrid and its parents, NC 25P and NC 30P. Hort-Science 45: 824—825.

Gonzalez-Cendales Y, Catanzariti AM, Baker B, Mcgrath DJ, Jones DA (2015) Identification of *I-7* expands the repertoire of genes for resistance to *Fusarium* wilt in tomato to three resistance gene classes. Mol Plant Pathol.

Goode MJ, Sasser M (1980) Prevention-the key to controlling bacterial spot and bacterial speck of tomato. Plant Dis 64: 831–834.

Goodwin SB, Sujkowski LS, Fry WE (1995) Rapid evolution of pathogenicity within clonal lineage of the potato late blight disease fungus. Phytopathology 85: 669-676.

Grattidge RO, Brien RG (1982) Occurrence of a third race of *Fusarium* wilt of tomatoes in Queensland. Plant Dis 66: 165-166.

Grimault V, Prior P, Anais G (1995) A monogenic dominant resistance of tomato to bacterial wilt in Hawaii-7996 is associated with plant colonization by *Pseudomonas solanacearum*. J Phytopathol 143: 349-352.

Gul Z, Ahmed M, Khan ZU, Khan B, Iqbal M (2016) Evaluation of tomato lines against Septoria leaf spot under field conditions and its effect on fruit yield. Agricultural Sciences 7(04): 181.

Guo SC, Zhuang JR, Li Q, Dong, QC (2008) Investigation and integrative management of bacterial spot in Fei County. China Veg: 5. 60–61.

Gupta PK, Balyan HS, Sharma PC, Ramesh B (1996) Microsatellites in plants: a new class of molecular markers. Curr. Sci. 70: 45–54.

Hanson P, Lu SF, Wang JF, Chen W, Kenyon L, Tan CW, Tee KL, Wang YY, Hsu YC, Schafleitner R, Ledesma D (2016) Conventional and molecular marker-assisted selection and pyramiding of genes for multiple disease resistance in tomato. Scientia Horticulturae. 201: 346-354.

Hanson PM, Licardo O, Hanudin, Wang JF, Chen JT (1998) Diallel analysis of bacterial wilt resistance in tomato derived from different sources. Plant Disease. 82: 74-78.

Hedrick WP, Booth NO (1907) Mendalian characters in tomatoes. Proc Am Soc Hortic Sci 5: 19-24.

Hernandes R, Stamova L (1990) Resistance to *Leveillula taurica* in tomato lines from Costa Rica. Tomato Genetics Cooperative Report 40: 15.

Hernandez MM (1985) Comparison of the effects of three isolates of *Stemphylillm solani* Weber, obtained from different regions of the province of Havana on lines of tomato (*Lycopersicon esc!llentum* Mill). Cult Trop 7(3): 3-11.

Hutton SF, Scott JW (2014) *Ty-6*, a major begomovirus resistance gene located on chromosome 10. Rept Tomato Genet Coop 64: 14–18.

Hutton SF, Scott JW, Jones JB (2010a) Inheritance of resistance to bacterial spot race T4 from three tomato breeding lines with differing resistance backgrounds. J Am Soc Hortic Sci 135: 150–158.

Hutton SF, Scott JW, Yang W, Sim SC, Francis DM, Jones JB (2010b) Identification of QTL associated with resistance to bacterial spot race T4 in tomato. Theor Appl Genet. 121: 1275–1287.

Ishibashi K, Masuda K, Naito S, Meshi T, Ishikawa M (2007) An inhibitor of viral RNA replication is encoded by a plant resistance gene. Proc Natl Acad Sci USA *104:* 13833-13838.

Jain RK, Pappu HR, Krishanareddy M, Vani A (1998) Watermelon bud necrosis tospovirus is a distinct virus species belonging to serogroup IV. Arch. Virol 143: 16371644.

Ji Y, Schuster DJ, Scott JW (2007a) Ty-3, a begomovirus resistance locus near the Tomato yellow leaf curl virus resistance locus Ty-1 on chromosome 6 of tomato. Mol. Breed. 20: 271–284.

Ji Y, Scott J.W. 2009. A CAPS marker linked to the tomato gray leafspot (*Stemphyllium sp.*) resistance gene *Sm. Rep Tomato Genet Coop* 59: 29–31.

Ji Y, Scott JW, Hanson P, Graham E, Maxwell DP (2007b) "Sources of resistance, inheritance, and location of genetic loci conferring resistance to members of the tomato-infecting begomoviruses," in Tomato Yellow Leaf Curl Virus Disease: Management, Molecular Biology, Breeding for Resistance, ed. H. Czosnek (Dordrecht: Springer), 343–362.

Ji, Y., Scott, JW, Schuster DJ, Maxwell DP (2009) Molecular mapping of *Ty-4*, a new Tomato yellow leaf curl virus resistance locus on chromosome 3 of tomato. J. Am Soc Hortic Sci 134: 281–288.

Jiang YC, Zeng LF (1983) Preliminary report on identification of bacterial spot of tomato. Yunnan Agric Sci Technol 1: 29 48.

Johanson A, Thurston HD (1990) The effect of cultivar maturity on the resistance of potato to early blight caused by *Alternaria solani*. Am Potato J 67: 615–623.

Jones JB (1991) Bacterial speck. In compendium of tomato diseases, ed. JB Jones, JP Jones, RE Stall, TA Zitter, pp. 26–27. St. Paul, MN: APS Press.

Jones JB, Lacy GH, Bouza H, Minsavage GV, Stall RE, Schaad NW 2005. Bacterial spot –Worldwide distribution, im- portance and review. Acta Hortic 695: 27–33.

Jones JB, Lacy GH, Bouzar H, Stall RE, Schaad NW (2004) Reclassification of the xanthomonads associated with bacterial spot disease of tomato and pepper. Syst Appl Microbiol 27: 755–762.

Jones JB, Zitter TA, Momol TM, Miller SA 2014. Compendium of Tomato Diseases and Pests, Second ed. APS Press, Minnesota, USA.

Jones RA, Millet A, Gainnini C (1989) A greenhouse screening method for *Pyrenochaeta lycopersici* resistance in tomato. Euphytica 40: 187-191

Jung J, Kim HJ, Lee JM, Oh CS, Lee HJ, Yeam I (2015) Gene based molecular marker system for multiple disease resistances in tomato against Tomato yellow leaf curl virus, late blight, and verticillium wilt. Euphytica *205*: 599-613.

Jung J, Kim HJ, Lee JM, Oh CS, Lee HJ, Yeam I (2015) Gene-based molecular marker system for multiple disease resistances in tomato against *Tomato yellow leaf curl virus*, late blight, and verticillium wilt. Euphytica 205: 599-613.

Kalloo G (1986) Tomato (*Lycopersicon esculentum* Miller). Allied Publishers Pvt, New Delhi, 470 pp.

Kanwar JS, Kerr EA, Harney PM (1980a) Linkage of *Cft* to *Cftl* genes for resistance to tomato leaf mold (*C. fulvum* CKe). Rep Tomato Genet Coop 30: 20-21.

Kanwar JS, Kerr EA, Harney PM (1980b) Linkage of *Cft2* to *Cf24* genes for resistance to tomato leaf mold (*C. fulvum* CKe). Rep Tomato Genet Coop 30: 22-23.

Kawchuk LM, Hachey J, Lynch DR, Kulcsar F, van Rooijen G, Waterer DR, Robertson A, Kokko E, Byers R, Howard RJ, Fischer R, Prufer D. (2001) Tomato *Ve* disease resistance genes encode cell surface-like receptors. Proc Natl Acad Sci USA 98: 6511-6515.

Kerr E A (1977) Problems and complications of developing new greenhouse varieties. In: Proc Can Mid- west greenhouse vegetable grower, Ontario, Canada pp 45-47.

Kerr EA, Bailey DL (1964) Resistance to *Cladosporium fulvum* Cke. obtained from wild species of tomato. Can J Bot 42: 1541-1542.

Kerr EA, Kerr E, Patrick ZA, Potter JW (1980) Linkage relation of resistance to *Cladosporium* leafmould (*Cf2*) and root-knot nematodes (*Mi*) in tomato and a new gene for leaf mould resistance (*Cf11*). Can J Genet Cytol22: 183-186.

Kerr EA, Patrick ZA, Bailey DL (1971) Resistance in tomato species to new races of leaf mold (*Cladosporium fulvum* Cke). Hort Res 11: 84-92.

Kim BY, Hwang IS, Lee HJ, Lee JM, Seo EY, Choi D, Oh CS (2018) Identification of a molecular marker tightly linked to bacterial wilt resistance in tomato by genome-wide SNP analysis. Theor Appl Genet.

Kim MJ, Mutschler MA (2005) Transfer to processing tomato and characterization of late blight resistance derived from *Solanum pimpinellifolium* L. L3708. J Am Soc Hor Sci 130: 877–884.

Kuklev MY, Fesenko IA, Karlov GI (2009) Development of a CAPS marker for the Verticillium wilt resistance in tomatoes.Russian journal of genetics 45(5): 575-579.

Kunkalikar SR, Sudarsana P, Arun BM, Rajagopalan P, Chen TC, Yeh SD, Naidu RA, Zehr UB, Ravi KS (2011) Importance and genetic diversity of vegetable-infecting tospoviruses in India. Phytopathology 101: 367376.

Kurozawa C, Balmer E (1977) Response of national and imported cultivars and wild species of tomato to *Septoria lycopersici* Spe.g. Summa. Phytopathologica 3: 135-141.

Lanfermeijer FC, Dijkhuis J, Sturre MJG, de Haan P, Hille J (2003) Cloning and characterization of the durable tomato mosaic virus resistance gene *Tm*-22 from *Lycopersicon esculentum*. Plant Molecular Biology 52: 1037–1049.

Lanfermeijer FC, Warmink J, Hille J (2005). The products of the broken Tm-2 and the durable *Tm-2*(2) resistance genes from tomato differ in four amino acids. J Exp Bot *56*: 2925-2933.

Lapidot M, Karniel U, Gelbart D, Fogel D, Evenor D, Kutsher Y (2015) A novel route controlling begomovirus resistance by the messenger RNA surveillance factor pelota. PLoS Genet.

Laterrot H (1978) Resistance aux maladies. B. *Pyrenochaeta lycopersici*. Rapp Activ Stn Amelion Plantes Moraic 1977-1978: 102-103.

Laterrot H (1986) Race 2.5.9, a new race of *Cladosporium fulvum* (*Fulvia fulva*) and sources of resistance in tomato. Neth Plant Pathol 92: 305-307.

Laterrot H (1994) NILs of tomato except for *Ph-2* gene. Tomato Coop. Genet. Rprt 44: 20–21.

Lawson VF, Summers WL (1982) Screening wild Lycopersicon for resistance against Pseudomonas tomato and *Xanthomonas uesicatoria*. HortScience 17, 503.

Li C, Jin Q, Peng G (1997) Pathogen identification of tomato bacterial scab in Xinjiang. China Veg: 4 4–6.

Li HT, Zou QD, Lv SW, Mu X, Xu WK (2001) Study on inheritance to bacterial wilt in tomato. Liaoning Agr Sci: 5 1-4.

Li MY (1991) Identification and control of tomato diseases. Vegetables: 3, 26–27.

Li YL, Zhang W, Zhang XY, Li DZ (2012) Cross breeding of canker-resistance varieties for tomato and identification by SSR molecular markers. Hunan Agric Sci: 13 13–15.

Liabeuf D (2016) Development of Processing Tomato Lines Resistant to *Xanthomonasgardneri*: From Screening to Breeding [Ph.D. Dissertation]. The Ohio State University, Columbus, Ohio, USA.

Liabeuf D, Francis DM, Sim SC (2015) Screening cultivated and wild tomato germplasm for resistance to *Xanthomonas gardneri*. Acta Hortic 1069: 65–70.

Lim GTT, Wang GP, Hemming MN, Basuki S, McGrath DJ, Carroll BJ, Jones DA (2006) Mapping the *I-3* gene for resistance to Fusarium wilt in tomato: application of an *I-3* marker in tomato improvement and progress towards the cloning of *I-3*. Australas. Plant Pathol. 35: 671-680.

Lim GTT, Wang GP, Hemming MN, McGrath DJ, Jones DA (2008) High resolution genetic and physical mapping of the *I-3* region of tomato chromosome 7 reveals almost continuous microsynteny with grape chromosome 12 but interspersed microsynteny with duplications on *Arabidopsis* chromosomes 1, 2 and 3. Theor Appl Genet 118: 57-75.

Lindhout P, Pet G, Van der Beek H (1994) Screening wild *Lycopersicon* species for resistance to powdery mildew (*Oidium lycopersicum*). Euphytica 72: 43-49.

Ling KS, Scott JW (2007) Sources of resistance to Pepino mosaic virus in tomato accessions. Plant disease 91(6), 749-753.

Locke SB (1949) Resistance to early blight and septoria leaf spot in the genus *Lycopersicon*. Phytopathology 39: 824-836.

Lodha PC (1977) Reaction of some tomato cultivars to culture filtrate of *Alternaria solani* (Ed et Mat) Jones et grout. Phytopathol Mediterr 16: 36-37.

Lukyanenko AN (1991) Disease resistance in tomato. *In*: *Genetic Improvement of Tomato*. (ed. G. Kalloo), pp. 99–119. Springer-Verlag, Berlin Heidelberg.

Maiero M, Ng TJ, Barksdale TH (1989) Combining ability estimates for early blight resistance in tomato. J Am Soc Hort Sci 114: 118–121

Mandal B, Jain RK, Krishnareddy M, Kumar NKK, Ravi KS, Pappu HR (2012). Emerging problems of Tospoviruses (Buniyaviridae) and their management in the Indian Subcontinent. Plant Dis 96, 468479.

Mangin B, Thoquet P, Olivier J, Grimsley N H (1999) Temporal and multiple quantitative trait loci analyses of resistance to bacterial wilt in tomato permit the resolution of linked loci. Genetics 151: 1165-1172.

Manjunatha L (2008). Studies on bud blight disease of tomato caused by groundnut bud necrosis virus (Doctoral dissertation, UAS, Dharwad).

Mansilla-Cordova PJ, Bampi D, Rondinel-Mendoza NV, Melo PCT, Lourencao AL, Rezendea, J. A. M. 2018. Screening tomato genotypes for resistance and tolerance to Tomato chlorosis virus. *Plant Pathology*, 67: 1231–1237; Doi: 10.1111/ppa.12826

Martin GB, Brommonschenkel SH, Chunwongse J, Frary A, Ganal, MW, Spivey R,Wu TY, Earle ED,Tanksley SD (1993) Map-based cloning of a protein-kinase gene conferring disease resistance in tomato. Science 262: 1432–1436.

Matýas Gonza ́lez-Arcos, Maria Esther de Noronha Fonseca, Ana Arruabarrena, Mirtes F. Lima, Miguel Michereff-Filho, Enrique Moriones, Rafael Fernandez-Munoz, Leonardo S. Boiteux (2018) Identification of genetic sources with attenuated Tomato chlorosis virus-induced symptoms in *Solanum* (section *Lycopersicon*) germplasm. Euphytica 214: 178.

Miao LX, Shou SY, Cai JY, Jiang F, Zhu ZJ, Li H B (2009) Identification of two AFLP markers linked to bacterial wilt resistance in tomato and conversion to SCAR markers. *Mol Biol Rep* 36: 479-486.

Miller AN, NgT, Barksdale TH (1983) A diallel analysis of the inheritance of resistance to tomato anthracnose caused by *Colletotrichum dematium.* Phytopathology 73: 968.

Milligan SB, Bodeau J, Yaghoobi J, Kaloshian I, Zabel P, Williamson VM (1998) The root knot nematode resistance gene *Mi* from tomato is a member of the leucine zipper, nucleotide binding, leucine-rich repeat family of plant genes. Plant Cell 10: 1307-1319.

Moore WD, Reynard GB (1945) Varietal resistance of tomato seedlings to the stem -lesion phase of *Alternaria solani.* Phytopathology 35: 933-935.

Moreau P, Thoquet P, Olivier J, Laterrot H, Grimsley N. (1998) Genetic mapping of *Ph-2*, a single locus controlling partial resistance to *Phytophthora infestans* in tomato. Mol. Plant Microbe Interact 11, 259-269.

Moriones E, Navas-Castillo J (2000) Tomato yellow leaf curl virus, an emerging virus complex causing epidemics worldwide. Virus Res. 71, 123–134.

Nowakowska M, Nowicki M, K³osiñska U, Maciorowski R, Kozik EU (2014) Appraisal of artificial screening techniques of tomato to accurately reflect field performance of the late blight resistance.PloS one 9(10).

Nowicki M, Foolad MR, Nowakowska M, Kozik EU (2012) Potato and tomato late blight caused by *Phytophthora infestans*: an overview of pathology and resistance breeding. Plant Dis. 96, 4—17.

Ohlson EW, Foolad MR (2015) Heritability of late blight resistance in tomato conferred by *Solanum pimpinellifolium* accession PI 224710. Plant Breeding 134(4): 461-467.

Ohnishi J, Yamaguchi H, Saito A (2016) Analysis of the Mild strain of tomato yellow leaf curl virus, which overcomes *Ty-2*. *Arch.Virol.* 161: 2207– 2217.

Oliveira RM, Dianese ÉC, Lima MF, Resende RO, Inoue-Nagata AK, Boiteux LS (2018) Sources of resistance to Potato virus Y and Pepper yellow mosaic virus in *Solanum* (section *Lycopersicon*) germplasm. European journal of plant pathology: 150(3): 691-699.

Opena RT, Green SK, Talekar NS, Chen JT (1989) Genetic improvement of tomato adaptability to the tropics: progress and future prospects. In: Tomato and Pepper Production in the Tropics. Proceedings of the International Symposium on Integrated Management Practices, S.K. Green (Sc. Ed.), T.D. Griggs and B.T. Mc Lean (Publ. Eds), AVRDC, Shanhua, Taiwan, Pp.71-85.

Palti J. (1988). The *Leveillula* mildews. The Botanical Review 54(4): 423–535.

Panpan Dong, Koeun Han, Muhammad Irfan Siddique, Jin-Kyung Kwon, Meiai Zhao, Fu Wang, Byoung-Cheorl Kang (2016). Gene-Based Markers for the Tomato Yellow Leaf Curl Virus Resistance Gene *Ty*-3.

Panthee DR, Foolad MR. (2012) A reexamination of molecular markers for use in marker-assisted breeding in tomato. Euphytica 184: 165-179.

Panthee DR, Gardner RG (2010) 'Mountain Merit': a late blight-resistant large-fruited tomato hybrid. HortScience 45: 1547— 1548.

Panthee DR, Gardner RG, Ibrahem R, Anderson, C (2015) Molecular markers associated with *Ph-3* gene conferring late blight resistance in tomato. American Journal of Plant Sciences 6(13): 2144.

Parniske M, Hammond-Kosack KE, Golstein C, Thomas CM, Jones DA, Harrison K, Wulff BB, Jones JD (1997) Novel disease resistance specificities result from sequence exchange between tandemly repeated genes at the Cf-4/9 locus of tomato. Cell 91: 821-832.

Pedley KF, Martin GB (2003) Molecular basis of *Pto* -mediated resistance to bacterial speck disease in tomato. Annu Rev Phytopathol 41: 215–243.

Pei C, Wang H, Zhang J, Wang Y, Francis DM, Yang W. (2012) Fine mapping and analysis of a candidate gene in tomato accession PI128216 conferring hypersensitive resistance to bacterial spot race T3. Theor Appl Genet 124: 533-542.

Peirce LC (1971) Linkage tests with *Ph* conditioning resistance to race 0, *Phytophthora infestans*. Tomato Genet Co-op Rep 21: 30.

Pelham J (1966) Resistance in tomato to tobacco mosaic virus. *Euphytica* 15, 258–267.

Pereira-Carvalho RC, Boiteux LS, Fonseca MEN, Díaz-Pendón JA, Moriones E, Fernández-Muñoz R (2010). Multiple resistance to *Meloidogyne spp*. and bipartite and monopartite Begomovirus spp. in wild *Solanum* (*Lycopersicon*) accessions. Plant Dis 94: 179–185.

Pilowsky M, Zutra D (1982) Screening wild tomatoes for resistance to bacterial speck pathogen (*Pseudomonas tomato*). Plant Disease 66: 46-4 7.

Pitblado RE, Kerr EA (1980) Resistance to bacterial speck (*Pseudomonas tomato*) in tomato. Acta Horticulturae 100: 379-382

Poysa V, Tu JC (1996) Response of cultivars and breeding lines of *Lycopersicon* spp., to *Alternaria solani*. Can. Plant Dis. Surv 76: 5–8.

Rattan RS, Saini SS (1979) Inheritance of resistance to fruit rot (*Phytophthora parasitica* Dast.) in tomato (*Lycopersicon escillennun* Mill.). Euphytica 28: 315-317.

Reddy DVR, Ratna AS, Sudarshana MR, Kiran Kumar I (1992) Serological relationships and purification of bud necrosis virus, tospovirus occurring in peanut (*Arachis hypogaea* L.) in India. Ann Appl Biol 120: 279-286.

Ren Z, You Z, Munir S, Zhang Y, Li H, Zhang J, Wang T, Zheng W,Ye Z (2019) Development of a highly specific co-dominant marker for genotyping the *Ph-3* (tomato late blight resistance) locus by comparing cultivated and wild ancestor species. Molecular breeding 39(3): 45.

Renukadevi P, Nagendran K, Nakkeeran S, Karthikeyan G, Jawaharlal M, Alice D, Pappu HR (2015) First report of Tomato spotted wilt virus infection of chrysanthemum in India. Plant Disease 99(8): 1190-1190.

Rich AE, Richards MC, Meader EM (1962) Rockingham, a new tomato variety resistant to late blight. Plant Dis Rep 46: 130.

Robert VJ M, West MAL, Inai S, Caines A, Arntzen L, Smith JK, St Clair DA (2001) Marker-assisted introgression of blackmold resistance QTL alleles from wild *Lycopersicon cheesmanii* to cultivated tomato (*L. esculentum*) and evaluation of QTL phenotypic effects. Mol Breeding 8: 217–233.

Rossi M, Goggin FL, Milligan SB, Kaloshian I, Ullman DE, Williamson VM (1998) The nematode resistance gene *Mi* of tomato confers resistance against the potato aphid. *Proc Natl Acad Sci USA*, 95: 9750-9754. Plant Disease 66: 46-4 7.

Rotem J (1994) The genus *Alternaria* biology, epidemiology, and pathogenicity, 1st edn. APS, St. Paul, MN, pp 48–203.

Sadashiva AT, Hanson P, Reddy MK, Ravishankar KV, Prasad M, Prasanna HC, Reddy KM, Singh TH, Saritha RK, Hussain Z, Mythili JB, Kaushal A (2017) Breeding tomato (*Solanum lycopersicum* L.) for resistance to biotic and abiotic stresses.

Sain SK, Chadha ML (2016) "Response of some wild species of tomato against Peanut bud necrosis virus under open-field conditions." Indian Journal of Horticulture 73(1): 60-65.

Sasser JN (1989) Plant parasitic nematodes. "The Farmer's Hidden Enemy". Published by North Carolina State University Raleigh, North Carolina, pp.115.

Satyanaryana T, Mitchell SE, ReddyDVR, Brown S, Kresovich S, Jarret R, Naidu RA, Damski JW (1996) Peanut bud necrosis tospovirus S RNA: Complete nucleotide sequence, genome organization and homology to other tospoviruses. Arch. Virol. 141: 8598.

Schaible LW (1962) Fruit setting responses of tomato to high night temperature. In: Proc Plant Sci Symp Canden NJ, USA, pp 229.

Schornack S, Ballvora A, Gürlebeck D, Peart J, Baulcombe D, Ganal M, Baker B, Bonas U, Lahaye T. (2004) The tomato resistance protein Bs4 is a predicted non-nuclear TIR-NB-LRR protein that mediates defense responses to severely truncated derivatives of AvrBs4 and overexpressed AvrBs3. Plant J. 37: 46-60.

Scott JW, Francis DM, Miller SA, Somodi GC, Jones JB (2003) Tomato bacterial spot resistance derived from PI 114490, inheritance of resistance to race T2 and relationship across three pathogen races. J Am Soc Hortic Sci 128: 698–703.

Scott JW, Hutton SF, Shekasteband R, Sim SC, Francis DM (2013) Identification of tomato bacterial spot race T1, T2, T3, T4, and *Xanthomonas gardneri* resistance QTLs derived from PI 114490 populations selected for race T4. Acta Hortic 1069: 53–58.

Scott JW, Somodi GC, Jones JB (1988) Bacterial spot resistance is not associated with bacterial wilt resistance in tomato. Proc Fla State Hort Soc 101: 390-392.

Seah S, Williamson VM, Garcia B E, Mejia L, Salus M.S, Martin CT, Maxwell DP (2007) Evaluation of a co-dominant SCAR marker for detection of the *Mi-1* locus for resistance to root-knot nematode in tomato germplasm. Rep Tomato Genet Coop 57: 37-40.

Sen Y, Derwolf JV, Visser RGF, Heusden SV (2015) Bacterial canker of tomato: current knowledge of detection, management, resis- tance, and interactions. Plant Dis 99: 4–13.

Sharlach MD, ahlbeck D, Liu L, Chiu J, Jimenez-Gomez JM, Kimura S, Koenig D, Maloof JN, Sinha N, Minsavage GV, Jones JB, Stall RE, Staskawicz BJ (2013) Fine genetic mapping of RXopJ4, a bacterial spot disease resistance locus from *Solanum pennellii* LA716. Theor Appl Genet 126: 601–609.

Shrestha S, Kumar R, Behera TK, Sharma HK (2012) Inheritance of resistance to root-knot nematode (*Meloidogyne incognita* race 1) in tomato (*Solanum lycopersicum* L.) 'Pusa 120'. Journal of Horticultural Science and Biotechnology 87.3: 2110-216.

Sinden SL, Goth RW, O'Brien MJ (1972) Effect of potato alkaloids on the growth of *Alternaria solani* and their possible role as resistance factors in potatoes. Phytopathology 63: 303–307.

Singh SJ, Krishnareddy M (1996). Watermelon bud necrosis: A new Tospovirus disease. Acta Hort. 431: 6877.

Smart CD, Tanksley SD, Mayton H, Fry W. E. (2007) Resistance to *Phytophthora infestans* in *Lycopersicon pennellii*. Plant Dis 91: 1045-1049.

Smith EF (1910) A new tomato disease of economic importance. Science (New Series) 31: 794–796.

Smith PG (1944) Embryo culture of a tomato species hybrid. Proc Am Soc Hort Sci, 44: 413-416.

Sobir, Ohmori T, Murata M, Motoyoshi F (2000) Molecular characterization of the SCAR markers tightly linked to the *Tm-2* locus of the genus *Lycopersicon*. Theoretical and Applied Genetics 101: 64–69.

Stall RE, Jones JB, Minsavage GV (2009) Durability of resistance in tomato and pepper to *Xanthomonas* causing bacterial spot. Annu Rev Phytopathol 47: 265–284.

Stall RE, Walter M (1965) Selection and inheritance of resistance on tomato to isolates of races 1 and 2 of the Fusarium wilt organism. Phytopathology 55: 1213-1215.

Stamova L, Yordanov M (1987) Resistance to *Leveillula taurica* (Lev.) Am Tomato Genetics Cooperative Report 37: 73.

Stamova L, Yordanov M, Stamova S (1984) Genes for resistance to *Cladosporium fulvum* Cke (*Fulvia fulva*) in *Solanum pennellii* Corell. In: A new era in tomato breeding. 9th Meet Eucarpia Tomato Working Group, 22-24 May, 1984, Wageningen, Neth, pp 75-80.

Stevens MA, Rick CM (1986) Geneties and Breeding. In: Atherton JG, Rudich I (eds) The tomato crop: a scientific basis for improvement. Chapman and Hall, London, 35-109.

Stevenson WR, Evans GE, Barksdale TH (1978) Evaluation of tomato breeding lines for resistance to fruit anthracnose. Plant Disease Rep 62: 937-940.

Su X, Zhu G, Huang Z, Wang X, Guo Y, Li B, Du Y, Yang W, Gao J (2019) Fine mapping and molecular marker development of the Sm gene conferring resistance to gray leaf spot (*Stemphylium* spp.) in tomato. Theoretical and Applied Genetics *132*(4): 871-882.

Sun FZ, Zhu H, Zheng JQ, Yi Q (1991) Occurrence of bacterial spot in Beijing and Datong areas. Plant Prot 17: 50–51.

Sun HJ, Liu XX, Li WH, Yang WC (2011a) Preliminary mapping of a gene in tomato accession LA1589 conferring resistance to race T3 of bacterial spot. J Agric Univ Hebei 34: 65–69.

Sun HJ, Wei JL, Zhang JY, Yang WC (2014) A comparison of disease severity measurements using image analysis and visual estimates using a category scale for genetic analysis of resistance to bacterial spot in tomato. Eur J Plant Pathol 139: 125–136.

Sun HJ, Zhang JY, Wang YY, Scott JW, Francis DM,Yang WC (2011b) QTL analysis of resistance to bacterial spot race *T3* in tomato. Acta Hortic Sin 38: 2297-2308.

Sun WY, Zhao WY, Wang YY, Pei CC, Yang WC (2011c). Natural variation of *Pto* and *Fen* genes and marker-assisted selection for resistance to bacterial speck in tomato. Agric Sci China 10: 827–837.

Thapa SP, Miyao EM, Davis RM, Coaker G (2015) Identification of QTLs controlling resistance to *Pseudomonas syringae* pv. tomato race 1 strains from the wild tomato, *Solanum habrochaites* LA1777. Theoretical and Applied Genetics 128(4): 681-692.

Thirthamallappa, Lohithaswa HC (2000) Genetics of resistance to early blight (*Alternaria solani* Sorauer) in tomato (*Lycopersicum esculentum* L.). Euphytica 113: 187–193.

Thomas JE, McGrath DJ (1988) Inheritance of resistance to potato virus Y in tomato. Australian Journal of Agricultural Research 39(3): 475 – 479.

Tomás DM, Cañizares MC, Abad J, Fernández-Muñoz R, Moriones E (2011) Resistance to Tomato yellow leaf curl virus accumulation in the tomato wild relative *Solanum habrochaites* associated with the C4 viral protein. Mol Plant Microbe Interact 24, 849–861.

Truong HTH, Choi H, Cho MC, Lee HE, Kim JH (2011) Use of *Cf-9* gene-based markers in marker-assisted selection to screen tomato cultivars with resistance to Cladosporium fulvum. Horticulture, Environment, and Biotechnology 52(2): 204.

Truong HTH, Kim S, Tran HN 2015. Development of a SCAR marker linked to bacterial wilt (*Ralstonia solanacearum*) resistance in tomato line Hawaii 7996 using bulked-segregant analysis. Hortic Environ Biotechnol 56.4: 506-515.

Tuberosa, R. (Eds.), Genomics-Assisted Crop Improvement, vol 1: Genomics Approaches and Platforms. Springer, Dordrecht, The Netherlands, pp. 1–12.

Turina M, Kormelink R, Resende RO (2016) Resistance to tospoviruses in vegetable crops: epidemiological and molecular aspects. Annual review of phytopathology 54:.347-371.

Upinder Gill. John W. ScottReza ShekastebandEben OgundiwinCees SchuitDavid M. FrancisSung-Chur SimHugh SmithSamuel F. Hutton. 2019. *Ty-6*, a major begomovirus resistance gene on chromosome 10, is effective against Tomato yellow leaf curl virus and Tomato mottle virus. Theoretical and Applied Genetics 132(5): 1543–1554.

Vakalounakis DJ (1991) Control of early blight of greenhouse tomato, caused by *Alternaria solani*, by inhibiting sporulation with ultraviolet-absorbing vinyl film. Plant Dis 75: 795–797.

Veliev RV, Sviridova IA (1987) Breeding tomato for resistance to corky root. Sel Semenovod USSR 2: 26-27

Venkata Ramana C, Venkata Rao P, Prasada Rao RDVJ, Kumar SS, Reddy IP and Reddy YN. (2010) Genetic analysis for Peanut bud necrosis virus (PBNV) resistance in tomato (*Lycopersicon esculentum* Mill.). In *III International Symposium on Tomato Diseases* 914: 459-463.

Veremis JC, Roberts PA (1996) Differentiation of *Meloidogyne incognita* and *M. arenaria* novel resistance phenotypes in *Lycopersicon peruvianum* and derived bridge-lines. Theor Appl Genet 93: 960-967.

Verlaan MG, HuttonSF, Ibrahem RM, Kormelink R,Visser RGF, Scott JW (2013). The tomato yellow leaf curl virus resistance genes *Ty-1* and *Ty-3* are allelic and code for DFDGD-class RNA-dependent RNA polymerases. PLoS Genet.

Verlaan MG, Szinay D, Hutton SF, deJong H, Kormelink R, Visser RGF (2011). Chromosomal rearrangements between tomato and *Solanum chilense* hamper mapping and breeding of the TYLCV resistance gene *Ty-1*. Plant J 68: 1093–1103.

Vidavski F, Czosnek H, Gazit S, Levy D, Lapidot M (2008) Pyramiding of genes conferring resistance to Tomato yellow leaf curl virus from dierent wild tomato species. Plant Breed 127: 625–631.

Villafana RT, Ramdass AC Rampersad SN (2018) First report of *Colletotrichum truncatum* causing anthracnose in tomato fruit in Trinidad. Plant disease 102(9): 1857-1857.

Vloutoglou I (1999) Evaluation of tomato cultivars and hybrids for resistance to *Alternaria solani* infection. Tests Agrochem Cultiv 20: 48–49.

Wang CX, Ying CQ 2008 Occurrence pattern and prevention methods for bacterial spot of tomato. Jilin Veg 4: 50–51.

Wang GP, Lin MB, Wu DH. (2004) Classical and molecular genetics of bacterial wilt resistance in tomato. Acta Hortic Sin 31: 401–407.

Wang H, Hutton SF, Robbins MD, Sim SC, Scott JW, Yang WC, Jones JB, Francis DM (2011) Molecular mapping of hypersensitive resistance from tomato cv. Hawaii 7981 to *Xanthomonas perforans* race T3. Phytopathology 101: 1217–1223.

Wang JF, Hanson P, Barnes JA (1998) Worldwide evaluation of an international set of resistance sources to bacterial wilt in tomato. In: P. Prior, C. Allen and J. Elphinstone (eds.), Bacterial Wilt Disease: Molecular and Ecological Aspects, Springer-Verlag, Berlin, Pp. 269-275.

Wang JF, Ho FI, Truong HT H, Hugan SM, Balatero CH, Dittapongpitch V, Hidayati N. 2013. Identification of major QTLs associated with stable resistance of tomato cultivar 'Hawaii 7996' to *Ralstonia solanacearum*. Euphytica 190: 241-252.

Wang XH, Li GY, Ren YZ, Xue XW, Huang SF (2006a) Pathogen identification of bacterial spot of processing tomato in Xinjiang. Acta Agric Boreali occidentalis Sin, 15: 72–74.

Wang Y, Yang W, Zhang W (2013) Mapping of a Heat-Stable Gene for Resistance to Southern Root-Knot Nematode in *Solanum lycopersicum*. Plant Mol Biol Rep 31.2: 352-362.

Wang YY, Chai ALLi BJ (2015) Diversity of symptoms and integrative management of tomato bacterial speck. China Veg 11 82–84.

Watanabe LFM, Bello VH, De Marchi BR, Sartori MMP, Pavan MA, Krause-Sakate R (2018) Performance of Bemisia tabaci MEAM1 and *Trialeurodes vaporariorum* on Tomato chlorosis virus (ToCV) infected plants. J App Entomol 142, 1008–1015.

Watanabe Y, Kishibayashi N, Motoyoshi F, OkadaY (1987) Characterization of *Tm-1* gene action on replication of common isolates and a resistance-breaking isolate of TMV. Virology. *161*: 527-532.

Yadav OP, Dabbas MR (2012) Efficacy of fungicides in the management of early blight of tomato (Alternaria solani). International Journal of Plant Protection *5*(2): 413-416.

Yaghoobi J, Kaloshian I, Wen Y Williamson VM (1995) Mapping a new nematode resistance locus in *Lycopersicon peruvianum*. Theor Appl Gene 91: 457-464.

Yaghoobi J, Yates JL, Williamson VM (2005) Fine mapping of the nematode resistance gene *Mi-3* in *Solanum peruvianum* and construction of a *S. lycopersicum* DNA contig spanning the locus. Mol Genet Genomics 274: 60-69.

Yamaguchi H, Ohnishi J, Saito A, Ohyama A, Nunome T, Miyatake K, (2018). An NB-LRR gene, TYNBS1, is responsible for resistance mediated by the *Ty-2* Begomovirus resistance locus of tomato. Theor Appl Genet 131, 1345–1362.

Yan Z, Pérez-de-Castro A, Díez MJ, Hutton SF, Visser RGF, Wolters A-MA, Bai Y, Li J (2018) Resistance to tomato yellow leaf curl virus in tomato germplasm. Front Plant Sci 9: 1198.

Yang H, Zhao T, Jiang J, Wang S, Wang A, Li J, Xu, X (2017) Mapping and screening of the tomato *Stemphylium lycopersici* resistance gene, *Sm*, based on bulked segregant analysis in combination with genome resequencing. BMC Plant Biology, 17(1): 266.

Yang W, Francis DM (2005) Marker-assisted selection for combining resistance to bacterial spot and bacterial speck in tomato. J Amer Soc Hort Sci 130: 716-721.

Yang WC, Francis DM (2007) Genetics and breeding for resistance to bacterial diseases in tomato: prospects for marker assisted selection, in: Razdan, M.K., Mattoo, A.K. (Eds.), Genetic Improvement of Solanaceous Crops, 1. Tomato. Science Publishers, New Hamp- shire, USA: 379–419.

Yang X, Caro M, Hutton S F, Scott JW, Guo Y, Wang X (2014) Fine mapping of the tomato yellow leaf curl virus resistance gene *Ty-2* on chromosome 11 of tomato. Mol. Breed. 34: 749–760.

Yin XG, Wang XJ, Zhang Y, Pan GH, Yang QF (2005) Research progress on tomato bacterial wilt and resistance breeding in China. J Yunnan Agric Univ 20: 163–167.

Yordanov M, Stamova L, Stoyanova Z (1975). *Leveillula taurica* resistance in the tomato. Tomato Genetics Cooperative Report 25: 24.

Young ND, Tanksley SD (1989) RFLP analysis of the size of chromosomal segments retained around the *Tm-2* locus of tomato during backcross breeding. Theor Appl Genet 77: 353-359.

Yue SJ, Wu DH, Liang CY (1995) Studies on resistance heredity of bacterial wilt of tomato. J South China Agric Univ 16: 91-95.

Chapter 6

Speed Breeding in Vegetable Crops: A Smart Approach to Accelerate Cultivars Development

Shubham Singh, R. Ramesh Babu, C. Anil Kumar and Sandeep Yadav

ABSTRACT

The growing human population and changing environmental conditions significant concern for global food security, with the current improvement rate of several important crops which are inadequate to meet future demand. This slow improvement rate is attributed partly to the long generation time of crop plants. An alternate technique to counter this hurdle is the use of speed breeding technology that shortens the breeding cycle and accelerates the crop research through rapid generation advancement. It can be carried out in numerous ways, one of which involves extending the duration of plant's daily exposure to light, combined with early seed harvest, to cycle quickly from seed to seed, thereby reducing the generation time for some long-day or day-neutral crops. It can also achieve up to 6 generations per year in growth chambers and normal glasshouse conditions. The use of supplemental lighting using LEDs in a glasshouse environment allows rapid generation cycling through single seed descent (SSD) method and plant density can be scaled-up for large crop improvement programs. To improve agronomic traits associated with yield, quality, and resistance to biotic and abiotic stresses in crop plants, several conventional and molecular approaches coupled with speed breeding along with other method like selection, mutagenic breeding, somaclonal variations, whole-genome sequence-based approaches, physical maps, and functional genomics tools. However, recent advances in genome editing technology using programmable nucleases, clustered regularly interspaced short palindromic repeats (CRISPR), and CRISPR-associated (Cas) proteins have opened the door to a new plant breeding era.

Keywords: *Speed breeding, Genome editing, Food security, Conventional breeding, High throughput phenotyping and genotyping.*

Introduction

Over the next 30 years, the global human population is expected to grow by 25 per cent and reach 10 billion (Alahmad *et al.*, 2018). The scientist community is striving hard to develop improved cultivars having higher yielding and resistance against biotic and abiotic stress, using conventional breeding approaches. In the mid-1980s, NASA partnered with Utah State University to conduct joint research programme to search the possibility of growing rapid cycling wheat under constant light on space stations (Alahmad *et al.*, 2018). Inspired by NASA's work, researchers at the University of Queensland coined the term 'speed breeding' in 2003 for a set of improved methods to hasten wheat breeding.

Speed breeding can accelerate backcrossing and pyramiding of traits as well as transgenic pipelines (Chiurugwi *et al.*, 2019). The idea for growing of wheat in space inspired senior research fellow Dr Lee Hickey of the University of Queensland, Australia. The first spring wheat variety developed using speed breeding is 'DS Faraday', which was released in 2017 in Australia. Speed breeding activities can be carried out in an enclosed chamber with artificially provided LED light (*e.g.* halogen lamps) to provided PAR of 400-700 nm. The photoperiod of 22 hours with 2 hours of darkness in 24 hours of the diurnal cycle and temperature should be maintained accordingly and relative humidity 60-70 per cent also maintained during the entire life cycle (Hickey *et al.*, 2019). Alternatively cereals, speed breeding strategy also applicable in several vegetable crops like radish, Pea, Tomato (Tomato is susceptible to continuous light conditions but identification and introgression of continuous light tolerance gene (CAB-13) responsible to make them tolerant against continuous light and tend to increase productivity under continuous light), Amaranthus, Cassava, Potato, Brassica, Sugar beet and some other leafy vegetables crops (Bottesi *et al.*, 2016). Speed breeding is likely to reduce generation time for other crops like Tomato, Potato, Amaranthus (can produced eight- generations per year instead of two in the field), Pepper (early flowering and fruiting under continuous light. In mid-1980s, NASA partnered with Utah State University for initiated joint venture research for growing wheat in constant light condition in space research station which lead to emerge of new era "Speed breeding".

Principle

The principle behind speed breeding is to use optimum light intensity (photoperiod) light quality, optimum temperature (varied accordingly crop to crop), and daylight length control (22 h light, 22°C at day/17°C at night with high light intensity) (Ghosh and Watson, 2018) which leads to accelerate the rate of photosynthesis, stimulates early flowering, seed maturity, harvesting and ultimately shorten the generation time required for crop growth and development. Using speed breeding coupled with single seed descent method are more common for developing efficient breeding line and elite inbred line which can be exploited for the development of improved crop varieties in a shorter time, the speed breeding approach is more cheaper for developing inbred compared to the production of Di-haploids strategy. Speed breeding can be integrated with marker assisted selection (MAS) used for developing of target phenotypes (Alahmad *et al.*, 2018a).

Standard Protocol, Greenhouse Design and Set up-requirements for Speed Breeding

Vegetables are more sensitive for changing against macro and micro environment condition, instead a minor fluctuations in the environment tend to influence the growth and development of greater extent. To make successful approaches of speed breeding program a proper channel and set-up requirements

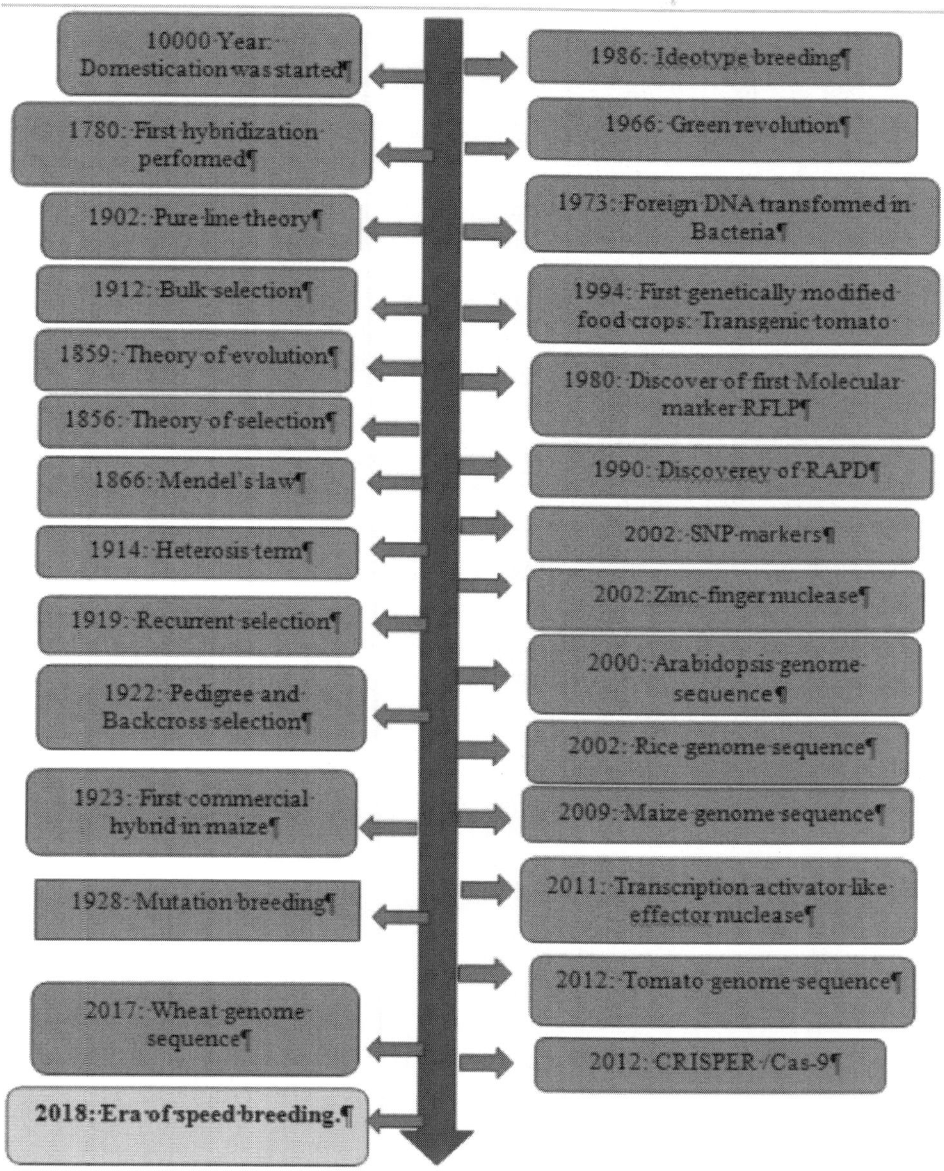

Figure 6.1. History and Evolution of Speed Breeding Technology.

are essential to adopted speed breeding in green-house structure, regulation on light (Photoperiod), temperature and humidity is apriory requirement (Amhar *et al.*, 2018). Photoperiod (Light) requirement differs plant species because different plants respond differently with varying wave-length emitted by lighting sources. Generally, light requirements covering PAR (400-700 nm) is more suitable for speed breeding with 22 hours of photoperiod. The optimum temperature should be maintained according to different plant species with specific growth stage (seedling germination, vegetative growth stage, flowering, fruiting, and fruit setting) with 60-70 per cent relative humidity (Bortesi *et al.*, 2016).

Methods Employed for Speed Breeding Program

(A) Controlled Environmental Chamber for Speed Breeding

This type of structures are equipped to a programmed regulation of up to 22-hour of photoperiod, with an optimum temperature of 22°C during the photoperiod, and about 17°C during the 2-hour dark period. An artificial photoperiod can be applied by a mixture of white LED bars (Valoya; 6 units per 3.67 m^2), far- red LED lamps (Valoya; 12 units per 3.67 m2) and metal HQI lamps (Valoya; 32 units per 3.67 m^2) and optimal light intensity should be adjusted to 360- 380 $\mu mol\ m^{-2}\ s^{-1}$ (highest value after ramping) at bench height, where pots is planted on the bench. Generally, wheat, barley, amaranthus and pea are suited to model plant for speed breeding under controlled environmental conditions. Available literatures are insufficient to elaborate the application of this approach in vegetable crops, but several attempts have been made to standardize the protocols in vegetables for application of speed breeding is underway.

(B) Glasshouse Speed Breeding Conditions

A temperature-controlled greenhouse is fitted and maintained at 17/22°C optimum temperature regime, with sodium vapour lamps and 12 hours of turnover and 22-hour of photoperiod (Ghosh and Watson 2018). Light intensity is maintained as 440-650 $\mu mol\ m^{-2}\ s^{-1}$ at adult plant height (approximately 45 cm above the bench height).

(C) Low-cost Homemade Growth Rooms for Speed Breeding

A homemade structure about 3 m x 3 m x 3 m with insulated sandwich panelling and fitted lighting equipment about 7 LB-8 LED light-boxes (1 light-box per 0.65 m^2) from Grow Candy (www.growcandy.com). Light quantity and quality about PAR at bench height ranged from 210-260 $\mu mol\ m^{-2}\ s^{-1}$and at 50 cm above the pot from 340-590 $\mu mol\ m^{-2}\ s^{-1}$. The lights were situated at a height of 140 cm above the bench. In this type of structure the room can accommodate 90 pots of 8″ diameter and 5 L volume are maintained at this kind of structure.

Application of Speed Breeding for Crop Improvement

Speed breeding is reduce generation time and help for achieving rapid genetic gain in several crops like such as sunflower (*Helianthus annuus*), pepper (*Capsicum annuum*), and radish (*Raphanus sativus*) (Anon 2018), Amaranthus (Bertesi *et al.*,

2016), model grass *Brachypodium distachyon* and other orphans crops were grown in a controlled environment room with extended photoperiod (22 hours light/2 hours dark). Orphan food crops are comparatively under-exploited or under-utilised plants and are reservoirs of genes to combat biotic and abiotic challenges such as climate change, diseases and pests threaten global food crops (Bertesi *et al.*, 2016, Bugbee *et al.*, 1997). Several reviews are published regarding the potential and used of speed breeding for crop improvement. In this review we discuss the application of speed breeding in several vegetables such as root crops (Radish), bulb crops (Onion) leafy vegetables (Amaranthus, Spinach, Lettuce *etc.*) and other minor vegetable crops. Development of speed breeding protocol has been reported in some vegetable crops which have been illustrated.

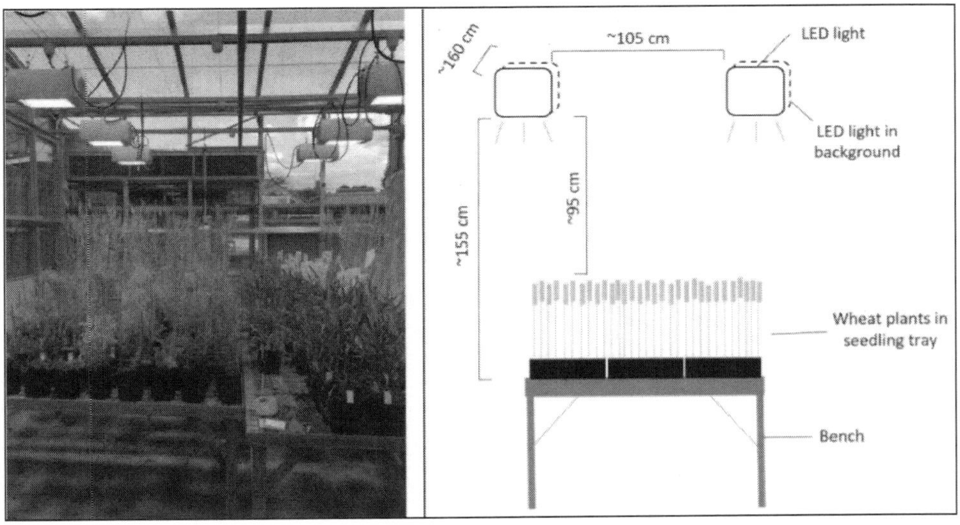

Figure 6.2. Layout of the Glasshouse Used for Speed Breeding at The University of Queensland, Australia (*Source*: Shreya Ghosh *et al.*, 2018).

Pea and Other Legume Crops

There are certaing speed breeding protocol have been developed in various legumes crop like lupin (Lupinus *sp.*) (Bugbee and Koerner, 1997), chickpea (*Cicer arietinum*) (Samineni *et al.*, 2020, Dev and Batia, 2019) subterranean clover (*Trifolium subterraneum*), lentil (*Lens culinaris*) and broad beans (*Vicia faba*) (Mobini 2015) and Pea (*Pisum sativum*) (Ray *et al.*, 2012). Speed breeding protocol for pea is illustrated as example here under (Chari and Church 2017). Matured dry seeds are sterilized and left for germinating on wet paper and transferred in a flask containing silver : sand (50:50) mixture and placed at dark condition for 5 days. Germinated seedling are inoculated with *Rhizobium* mixture, growing shoot are passed through neck and base should be covered with black plastic bag, an optimum temperature 22°C is maintained during all growing season and light was supplied at 12 hours and 16 hours photoperiod in two different batches, respectively (Anon, 2019). Plants are grown at 12 hours photoperiod, mature seeds are harvested 8 week while grown at 16 hours light (Photoperiod) condition (Mobini *et al.*, 2015).

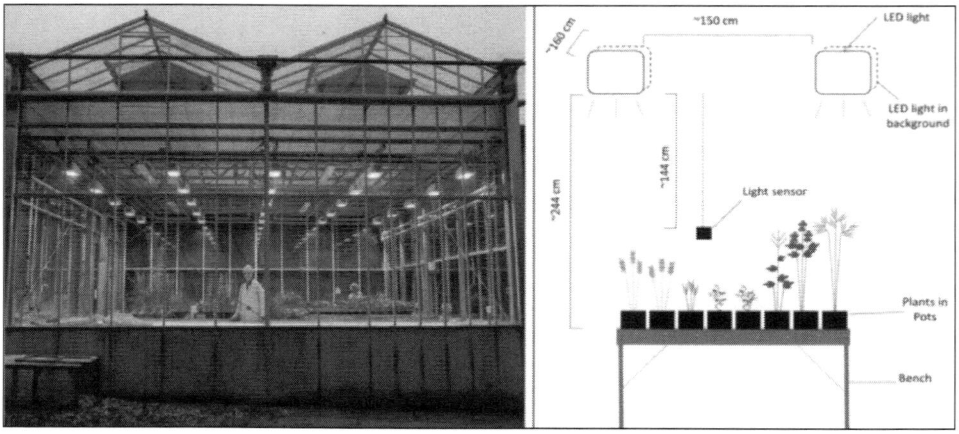

Figure 6.3. Layout of the Glasshouse Used for Speed Breeding at the John Innes Centre, UK. (*Source*: Shreya Ghosh *et al.,* 2018).

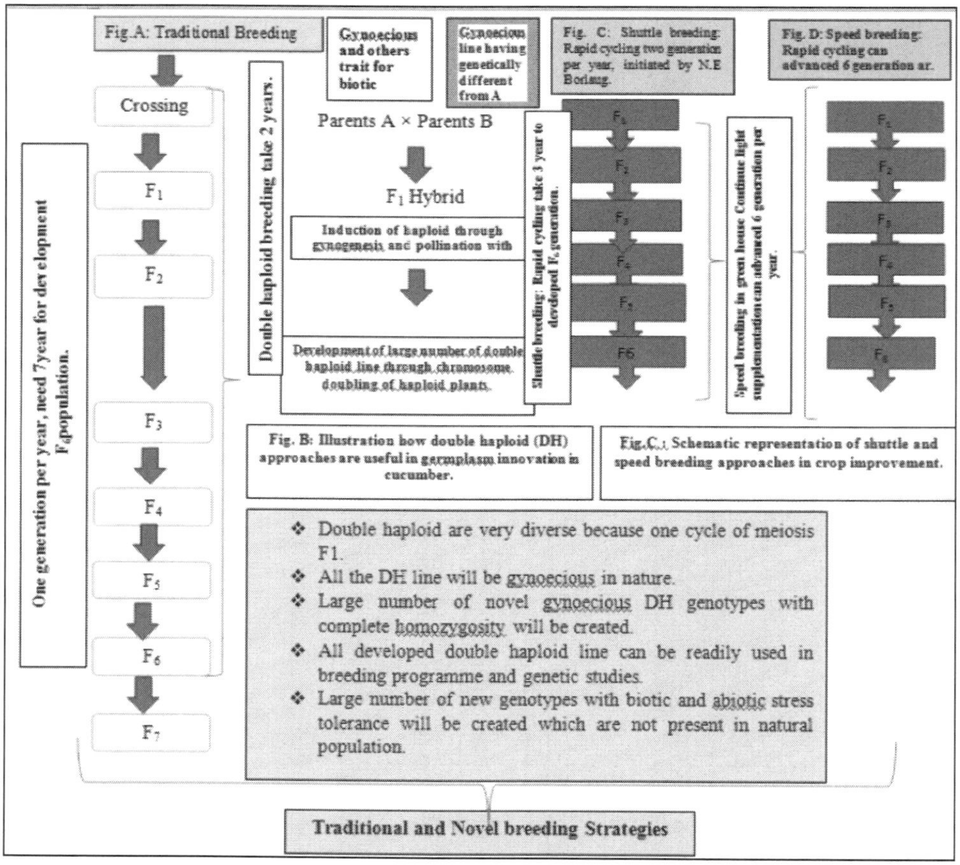

Figure 6.4. Schematic Representation of Traditional Breeding, Double Haploid Breeding, Shuttle Breeding and Speed Breeding for Rapid Gain of Desirable Trait.

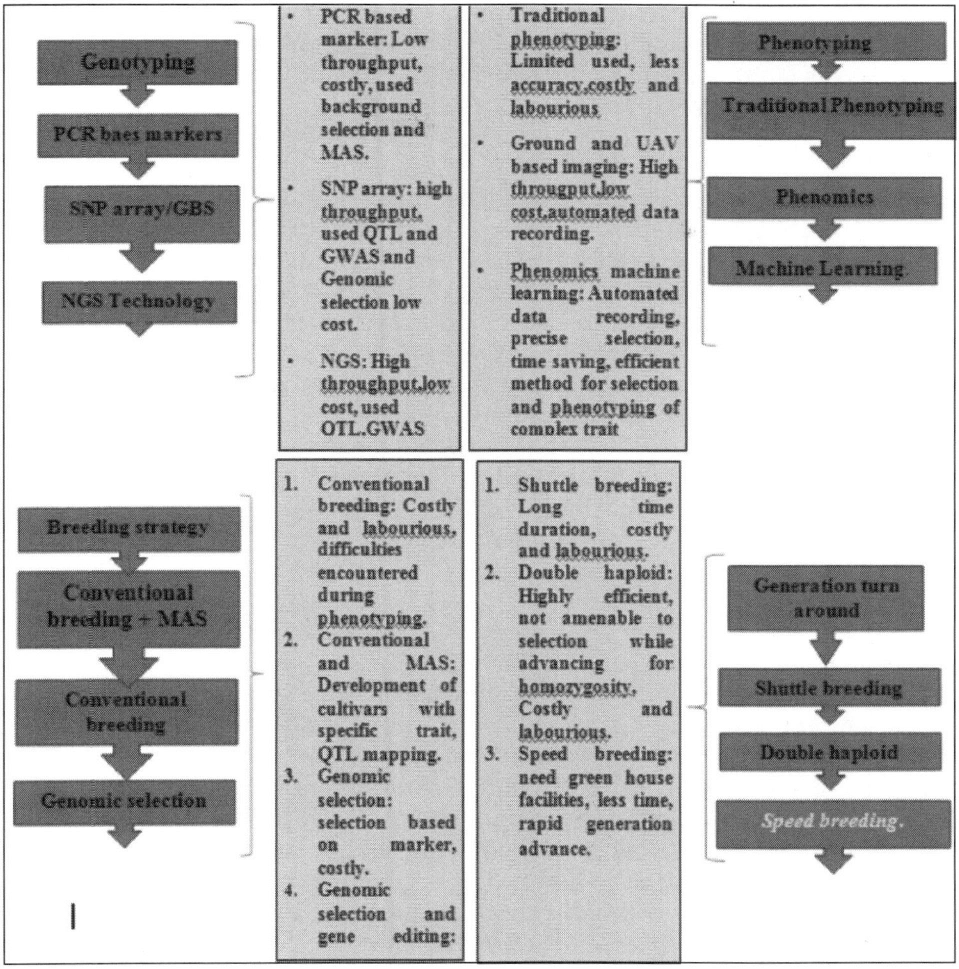

Figure 6.5. The Integrated Strategies Leading to a Common Decision Support Tool to Accelerate Genetic Gain Within the Context of New Technologies. features of individual aspects, such as genotyping, phenotyping, generation turn around, and breeding strategy are described in biplots, along with key advantages as follow (*Source*: Huihui Li and Awais Rasheed *et al.*, 2018).

Root and Tuber Vegetable Crops

There is no systematic research improvement for application of speed breeding and standard protocol for raising root and tuber vegetable crops (Chiurugwi *et al.*, 2019). Most of these are clonally propagated, and tuber vegetable are major source of calories requirement in certain developing African countries (Dey and Batia 2019), hence there is urge for development of vegetable cultivars within short time. Speed breeding facilitates make faster crosses and rapidly grows the progeny in preparation for the next round of breeding. Speed breeding protocols are yet to be developed for some tuber crops like cassava, sweet potato and root crops (Bartesi *et al.*, 2016, Liang *et al.*, 2017). Efforts have been made to shorten the time to flowering as well as flowering rate and yield predictability in Radish and carrot (Bugbee and Koerner, 1997).

Leafy Vegetable Crops

Leafy vegetables are annual, perennial and herbaceous growth habit (Sysoeva *et al.*, 2010), speed breeding protocol can accelerate earlier vegetative growth, early and synchronous flowering which facilitated fast breeding and varietal development (Bugbee and Keorner, 1997). The major breeding objective for application of speed breeding in leafy vegetable crops like amaranthus, spinach, lettuce *etc.* (Stetter *et al.*, 2016) is to improved pest and disease resistance and intra-specific uniformity in the harvested foliage improvement in foliage quality like enhancement of protein, Vitamins, other minerals and reduced anti-nutrient level in leafy vegetable crops (Blakeslee *et al.*, 1922). A significant achievement for application of speed breeding protocol has been developed in Amaranthus (Stetter *et al.*, 2016) to improve resistance to white rust (*Albugo bliti*) (Stetter *et al.*, 2016), affects the quality and quantity of foliage and grain yields (Ghosh and Watson, 2018).

Amaranthus

Ancient crops from the Americans such as quinoa (*Chenopodium quinoa* wild.) or Amaranth (*Amaranthus* spp. L.) (Bortesi *et al.*, 2016) are valuable addition to the human diet because of their high nutritional value (Chiurugwi *et al.*, 2019). Amaranthus is day neutral plant but some species are to be reported short day nature, artificially light supplementary at 16 hours photoperiod induced strong vegetative growth and after these plant are transferred at 8 hours + Short day photoperiod induced earlier flowering (Stetter *et al.*, 2016). Long day conditions lead to more vegetative growth, later flowering and more seeds, while short day conditions result in small early flower (Ghosh and Waston, 2018). Application of speed breeding promotes early flowering to advance breeding cycle. Supplementary of artificial light 16 hour photoperiod after crossing mature seeds could be harvested within 4 weeks. By employing these treatments and speed breeding 6-8 generations per year can be produced in amaranthus (Stetter *et al.*, 2016) instead 2 generation per year, which is comparable to the model plant *Arabidopsis thaliana* (Chiurugwi *et al.*, 2019)

Tomato

Tomato would be a good candidate for practical application of speed breeding techniques because most of indeterminate cultivars (Liang *et al.*, 2017, Chiurugwi *et al.*, 2019) are grown in greenhouse hence hastening earlier flowering, shorter vegetative growth. Also reported that increased nutritional and quality trait like mineral, vitamins and other trait in fruit can enhanced which are grown under greenhouse conditions (Dey and Batia 2019). Tomato is generally photoperiod sensitive nature hence introgression gene (*CAB-13*) to increased tolerant to long photoperiod enhance the efficiency of speed breeding concept (Sysoeva *et al.*, 2010) exposing a plant to 22 hours of artificial light and two hours of dark, significantly speed up the life cycle of production (Watson *et al.*, 2018). Tomato is very adapted to growing in the greenhouse and inbred lines could be generate within a short time (Bugbee and Koerner 1997). Combining speed breeding with other genomic tools like marker-assisted breeding (MAS and genomic selection), high throughput phenotyping could significantly improve the efficiency for breeding for heat tolerance in tomatoes (Liang *et al.*, 2017) which are represented in Figure 6.6.

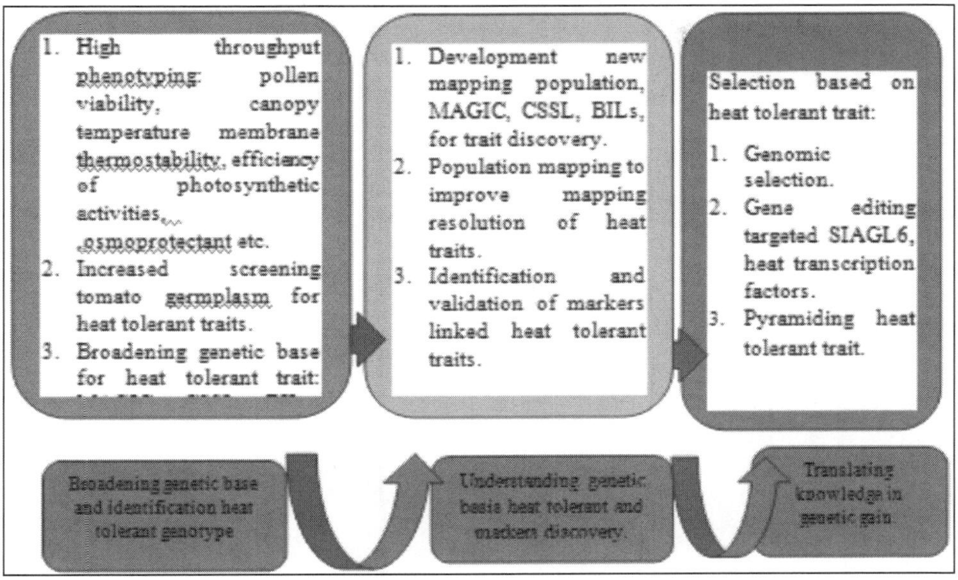

Figure 6.6. Discovery of Heat Tolerant Genetic Resource to Genetic Gain (*Source*: Mathieu Anatole T.A *et al.*, 2019).

Examples in other Vegetable Crops

Costes and Milhet (1970) reported early flowering and fruiting in pepper under continuous light compared 15 hours photoperiod (Liang *et al.*, 2017), and rapid senescence in onion and potato reported under continuous light condition reported by (Gestel *et al.*, 2005, Wheeler and Tibbitts *et al.* (1987).

Speed Breeding in Orphan Crops

Orphan crops have possess some various special traits (*e.g.*, low glycaemic indices and high concentrations of micronutrients)(Bortesi *et al.*, 2016), and they are considered major component for nutritional and food security in poor communities (Dey and Batia 2019). Orphan crops are also important for supporting affordable, sustainable and diverse farming systems because they require less water than most global staples and improve soils through nitrogen fixation and the incorporation of organic matter. Moreover, most orphan crops are hardy and can tolerate adverse conditions such as drought, frost, pests and disease (Bugbee and Koerner 1997). The major orphan crops include Finger millet, Sorghum, Cassava, Banana, Chickpea

(Samineni *et al.*, 2020), Rice, wheat (Anon, 2020) peanut and maize standard protocol for speed breeding have been developed for commercially utilization. The schematic representation speed breeding programme for peanut (Bortesi *et al.*, 2016) has been illustrated as Figure 6.7.

Speed Breeding for Fast Trait Stacking and Rapid Cycling than other Breeding Method

Crop breeding and advances in management practices have contributed substantially to the annual gain of 0.8–1.2 per cent in crop productivity (Amher *et*

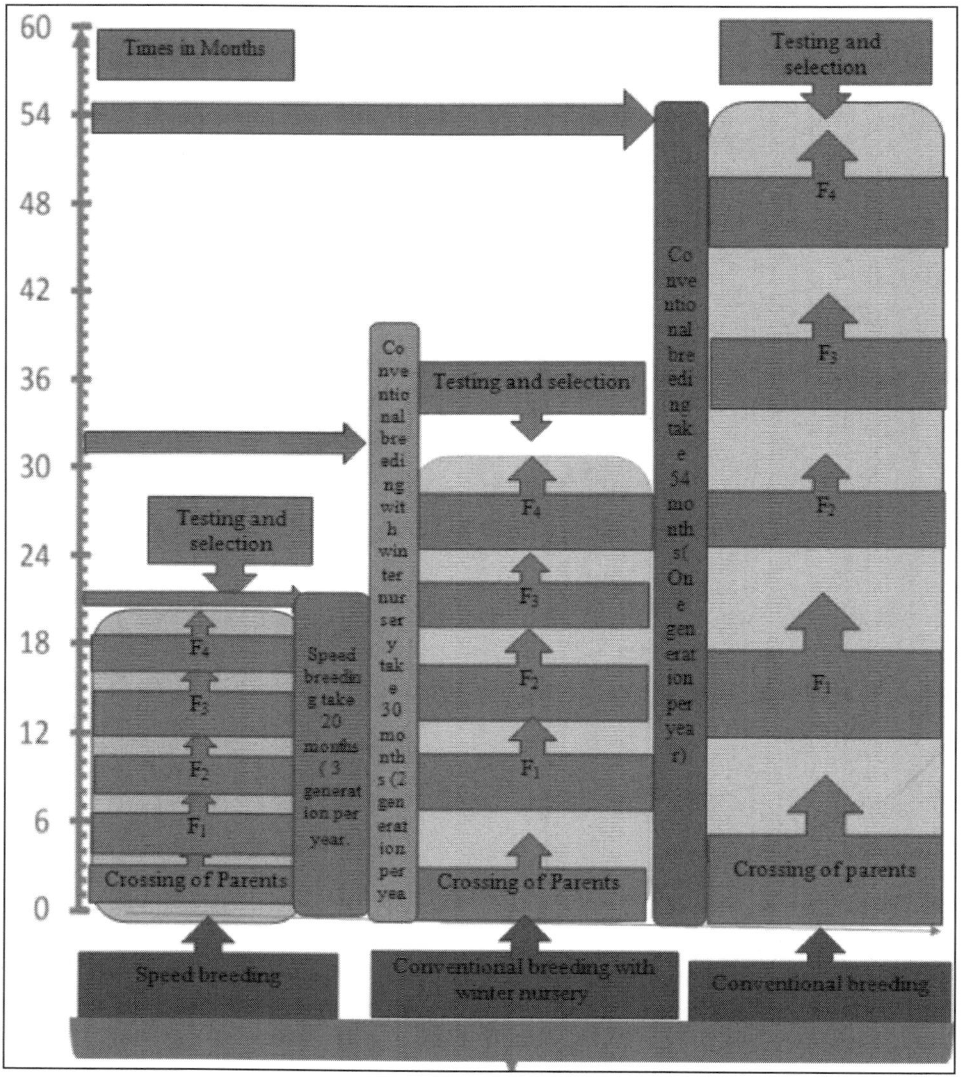

Figure 6.7. Schematic Comparison of Time Required to Develop Elite Lines from Different Breeding Method in Peanut.

Table 6.1. Rapid Generation Advanced Protocol for Speed Breeding in different Vegetable Crops (SD: Short day, LD: Long day)

Crops	*Greenhouse Generation Time (Days)*	*Photoperiod Response*	*Rapid Cycling Generation Time (Days)*	*Protocol Summary*
Soya bean	102-132	SD (Short day)	70	14 h light, 30°C day/25°C night, CO_2 (400-600 ppm), Increased crossing efficiency.
Rape seed	123	LD (Long day)	113	22h Light, 22°C day/17°C night, high-intensity PAR, Early seed harvest.
Cassava	365	LD (12-13 h for flowering)	-	SD and continuous application of fertilizer increased synchronous flowering.
Potato	138	LD or SD	-	Speed breeding with extended photoperiod developed (James Hutton institute)
Tomato	80	SD	-	Introgression of continuous light tolerance gene *CAB-13* to increased productivity under continuous light.
Sugar beet	Biennial	LD	-	-
Amaranthus	180 Days	Obligate SD.	-	8-h SD

al., 2020). The essence, genetic gain in a crop breeding program is determined by the following equation: ***DG ¼ i h sA=L;*** (Alahmad *et al.*, 2018, Gosh and Watson 2018) where *"i"* is the selection intensity, *"h"* is the square root of the heritability in the narrow sense, *"sA"* is the square root of the additive genetic variance, and *"L"* is the length of breeding cycle interval or generation (Anon, 2018). The evolution of "speed breeding" concept which provide a flexible platform for achieving rapid and fast genetic gain and trait stacking in several cereals and vegetable crops like wheat (*Triticum aestivum*)(Anon 2020), pea (*Pisum sativum*) (Watson *et al.*, 2018), amaranthus (*Amaranthus* spp.) (Stetter 2020), barley (*Hordeum vulgare*) (Bortesi *et al.*, 2016), Chickpea (*Cicer arietinum*) (Samineni *et al.*, 2020) and canola (*Brassica spp*) (Bortesi *et al.*, 2016), which can be achieved by growing 6-7 generation per year under continuous light supplement with fully controlled greenhouse structure (Ahmar *et al.*, 2020). The various integrated approaches with combine speed breeding, extensive genotyping and phenotyping by using modern tools, genomic selection, gene editing technology (Croser *et al.*, 2016) allow the selection of inbred, and increase efficiency of pre-breeding which accelerate the development of climatic-resilient cultivars in vegetable crops for future need (Andersson 2016). Genomic selection is useful tools to accelerate genetic gain with combining other low cost and high throughput genome sequencing technology, using low cost phenotyping (Chari and Church, 2017), SNP mining increased efficiency for rapid genetic gain. Now concentrated effort have been focus for whole genome sequencing, re-sequencing (Pan genome), Mut-map, mut-map+ RNA-seq and QTL-Seq (Chari and Church 2017) increased understanding the genetic lying under crop productivity, adoptability and higher yield (Aminul Islam, 2020). The CRISPER/Cas-9 tools is high throughput, multiplex genome editing technology regarded non-GMO approaches (Weller *et al.*, 2011), high throughput phenotyping tool are used to discovered haplotypes for various trait (early flowering, plant height, early flowering) in various vegetable crops are enabled for breeder to adopt new tools, breeding approaches and integrate with speed breeding, help for developing high yielding cultivars, better adoptability, better quality and climatic resilient cultivars in vegetable and other crops (Aminul Islam 2020). Generally 10-15 years are spent in releasing an elite variety using conventional breeding method but speed breeding program, growing up to six generations is possible per year for wheat, barley, chickpea and four generations per year for canola (Bugbee and Koerner 1997) and shorten the harvest time two to three times earlier than the conventional breeding approaches. Although it is a low cost program because it required minimum place and even operated with low cost LED bulbs for providing extended photoperiod for rapid cycling of plant growth (Sysoeva *et al.*, 2010), a glass house and larger breeding population of around 900 plants can be grown per meter square and efficient application of SSD approach (Bortesi *et al.*, 2016) used for selection. Thus signifies that a large breeding population can be handled by speed breeding by reducing their maintenance cost for developing inbreed lines (Crossa *et al.*, 2017).

Strategies to Reduce Length of Breeding Cycle

Plant breeders have strategies such as,

- ☆ Shuttle breeding
- ☆ Double haploid breeding
- ☆ Speed breeding

Speed Breeding Compares other Breeding Method

Embryo rescue techniques used for rapid cycling and fast generation advancement, employing rescue of embryo from immature seed and grow on culture media with artificially supplement all micro- and macro elements essential for growth and development with optimal temperature condition (Doudna *et al.*, 2014, Ray eta al., 2012). Using this approaches 4 generations per year can be achieved in lentil achieved 6-8 generation in pea and faba bean using this approaches (Mobini *et al.*, 2015). 8 generation in wheat (Ayenan *et al.*, 2019) and achieved 9 generation in barley (Bortesi *et al.*, 2016), 5 generation in soya bean reported by Roumet and Mohini *et al.* (1997) and 5-6 generation can be achieved in pea using embryo rescue techniques (Watson *et al.*, 2018).

Carbon di-oxide and their concentration in environment also influence varied plant growth activities like vegetative growth, earlier flowering, fruiting and pollen germination. Tanaka *et al.* (2016) reported rapid generation advancement of rice cultivars by elevated level and concentration of CO_2 combined with canopy thinning and slow root growth (34), followed by early harvest to hasten the generation times. Blakeslee *et al.* (1922) first identified naturally occurring haploids of *Datura stramonium* and subsequently natural haploids of many other plants were reported by various researchers in different vegetable crops (Andesson, 2016). The haploid embryo can arise from an egg cell (Gyanogenesis) (Ray 2012) or from a gametophytc cell other than the egg cell (apogamy) or a male gamete (androgenesis) (Dwivedi *et al.*, 2015). The observations in *Datura* were soon reproduced for tobacco both in France and Japan (Thomas *et al.*, 2003). This was followed quickly by success in the production of androgenic haploids in rice (Tanaka *et al.*, 2016). Sipra Guha (later Guha Mukherjee) shifted to anther culture of rice with Dr. M.S. Swaminathan at IARI-Pusa New Delhi (Doudna *et al.*, 2014, Dwivedi *et al.*, 2015). Double haploid are simply and highly attracted by plant breeder because it's simple and one step lead to producehomozygous line (Thomas *et al.*, 2003). Development of homozygous line in cucumber its take 1-2 year using double haploid approaches instead 6-7 generation (4-5 year) in conventional approaches (Crossa *et al.*, 2017). It has some constraint like highly expensive, technical expertise, restricted recombination to a single round of meiosis and labour intensive when used to handle a very large number of population (Anon, 2019).

Speed breeding approaches has its advantage like, generated advanced inbred line in 2-3 year which help to release commercial cultivars within 3-4 year in cucumber rather than it's take 7-8 year using conventional breeding approaches, rapid development of recombinant inbred line (RILs) within 4-5 year which are ideally suited for QTLs mapping, fine mapping, gene tagging and other molecular breeding like marker assisted selection (Chiurugwi *et al.*, 2019). Double haploid is a powerful tool for accelerate breeding and rapid development of completely homozygous stable line for difficult trait like for gynoecy and parthenocarpy without using marker assisted back-cross breeding (Thomos *et al.*, 2003). Muskmelon double haploid are produced by induction of haploid from F_1 and subsequent diploidization lead to produced double haploid in muskmelon (*Cucumis melo*) (Chiurugwi *et al.*, 2019).

CRISPR/Cas9 system is a high-throughput genome editing and GM trait are also incorporate breeding tools in speed breeding programme (Weller *et al.*, 2011). CRISPR/Cas-9 genome editing could be combine directly with speed breeding programme for efficient genetic modification for targeted trait, but still having some bottleneck such as, time consuming, and tissue culture protocol and well equipped laboratory facilities and desirable regeneration protocol are not well developed in many crops (Yang and Li, 2017).

Application and Constraint of Speed Breeding in Vegetable Improvement

Speed breeding can accelerate the generation advancement and develop the breeding population. Mobini *et al.* (2005) recorded and reported rapid development of mapping population in pea, mutation studied, fast-forwarding genetic trait, faster and better phenotyping. Marker- assisted selection coupled with speed breeding provides a flexible platform to rapid introgression of the desired gene for targeted trait having resistance to biotic and abiotic stress and pyramiding could help the development of multiple disease resistance cultivars in different cereals and vegetable crops. Speed breeding a powerful tool can lead accelerate genetic gain and rapid incorporation of desired trait. Speed breeding protocol have been standardized in vegetable soya bean, Amaranthus, developing and designing crop-specific lighting scheme that can accelerate early flowering and maturity and utilized for improving germination. Speed breeding can facilitated for introgression of simply inherited monogenic trait and tend to obtained up to five generation per year using speed breeding programme instead traditional method need 4-5 year, which lead double annual genetic gain compared any fast breeding programme. Further more several advanced breeding method such as marker assisted selection or genomic can be incorporated with speed breeding programme for developing of varieties.

Speed Breeding has some Limitations

Application of speed breeding protocol in various vegetable crops are not yet developed. (A) Most vegetables are long day and day-neutral plants, hence, continuous light can accelerate the genetic gain and breeding cycle. But, short day plants demands limited photoperiod for their growth and development which limits the application of speed breeding for improvement of short day vegetable crops. Speed breeding is more straight forward and easier to implement in long day and day-neutral types than short-day plant., but Lee Hickey *et al.* (2018) have been standardized speed breeding protocol for some short-day plants like millet, sorghum *etc.*, but not yet been developed in any vegetable crop. (B) Different phenotyping: speed breeding accelerates rapid cycling and shortens generation time. Further, early harvest of immature seed can interfere with phenotyping of some traits. (C) High initial investment: speed breeding programs are carried in enclosed glass house chambers which requires highly initial investment, purchasing lighting equipment and controlled capabilities and supplementary LED lighting provides more efficient power usage and reduced heat than other lighting types, such as SVLs. (D) Plants which are grown in an enclosed glass chamber with extended photoperiod show various physiological and toxicity symptoms like, chlorosis, necrosis and yellowing due to any micronutrient or heavy metal deficiency or excess.

Future Outlook of Speed Breeding in India

The experiments conduct by NASA (USA) to grow crop seeds in space inspired scientists in the University of Queensland and University of Sydney in Australia to develop a speed-breeding platform (Anon, 2018). One major advantage regarding quality and yield was reported that the crop grown in speed breeding glass house chamber are superior than normal glass house (Guha and Maheshwari, 1964). Farmer will be need to produced 60-80 per cent more food grain up to 2050 for feed 10 billion population (Guha and Maheshwari, 1966; Guha and Watson, 2018). In India several ICAR institute are looking to this forward but one institute (Gua and Maheshwari, 1964), ICAR-Indian Institute of Soybean Research (ICAR-IISR) (Guha and Maheshwari, 1966), Indore is extensively grown off-season nursery during months of (January–April) for rapid generation advancement of segregating material at the University of Agricultural Sciences, Bengaluru (Hickey *et al.*, 2019). This experiment accelerate and helps in reducing the duration for developing inbred lines for desirable trait which can used varietal development programme.

In addition, ICAR-IISR has greenhouse and polyhouse facility, which is being used for advancement of F_1 to F_2 and important individual plant selections during off-season (November–February) (Guha and Maheshwari, 1964). Speed breeding a new idea and tremendous potential in Indian condition combat various challenges like growing population, decrease land ratio, shortage of resources, shortage of cultivable land all these problems can be solve by adopting a nóvel breeding approaches which can accelerate rapid breeding cycling, development of high yielding cultivars at short duration to feed world second largest population country (Guha and Maheshwari 1966).

Future Thrust

To respond faster to the changing climate, evolving pathogens a breakthrough technology in genetics and plant breeding programme should be needed and speed breeding will definitely full fill that in future. NASA is looking at ways to provide astronauts with nutrients in a long-lasting, easily absorbed form freshly grown fresh fruits and vegetables through speed breeding technology. As it increases the breeding generation per year, a greater number of researches will be carried out and documented.

Conclusion

Recent advancements in genetic tools, breeding methods and development of NGS sequencing technology provides a flexible platform for sequencing, tagging, mapping and introgression of a gene for the desired trait at low cost, but it is time-consuming. Speed breeding coupled with other approaches tends to shorten the breeding cycle, generation advancement and accelerates the development of cultivars in a very short time. speed breeding and integration with other breeding technologies may lead rapid genetic gain, For example, tomato is sensitive to constant light, but researchers have identified a tomato gene (*CAB-13*) that make the plant to tolerate constant long photoperiod when transferred into a tomato cultivar grown under speed breeding conditions (20 per cent increase in fruit yield). Speed breeding as an alternative solution which can enable breeding in other recalcitrant crops, such as short-day species like maize and biennial species like sugar beet. It can be concluded that speed breeding combining other genetic tools like genome editing, Marker-assisted backcrossing, Transgenic approaches coupled with cost-effective genotyping and rapid phenotyping helps to develop homozygous inbred line followed by crossing will facilitate rapid cycling, genetic gain and rapid development of improved cultivars to researchers or plant breeder. Therefore speed breeding integrated with other multiple branches can be used to improve orphans, legumes, vegetables, leafy vegetables and other forage crops.

REFERENCES

Ahmar S, Rafaqat AG, Jung KH, Faheem A, Qasim MU, Zhou W (2020) Conventional and molecular techniques from simple breeding to speed breeding in crop plants: Recent advances and future outlook. Int. J. Mol. Sci. 21(7): 2590.

Alahmad S, Dinglasan E, Leung KM (2018) Speed breeding for multiple quantitative traits in durum wheat. Plant Methods 14: 36.

Alahmad S, Dinglasan E, Leung KM, Riaz A, Derbal N, Voss-Fels KP, Hickey LT (2018) Speed breeding for multiple quantitative traits in durum wheat. Plant Methods 14(1): 36.

Aminul Islam AKM (2020). Speed Breeding: An effective breeding technique for rapid generation advance in crop; EC Agriculture pp 9-11.

Amy Watson A, Ghosh S, Matthew JW *et al.* (2018). Speed breeding is a powerful tool to accelerate crop research and breeding. Nature Plants 4: 23–29.

Andersson M (2016) Turesson *gia plantarum*, 164(4), 378-384.

Annonimous (2018) https: //www.jic.ac.uk/news-and-events/news/2018/01/speed-breeding.

Annonimous (2019) https: //www.sciencedaily.com/releases/2018/01/180101144758.htm.

Annonimous (2020) Speed breeding for Indian Agriculture: a rapid method for development of new crop varieties.http: //www.newsweek.com/growing-plantsspeed-breeding-could-be-key-feed-worldsexploding-population 767450.

Arthur JM, Guthrie JD, Newell JM, Arthur JM. Guthrie JD, Newell JM. (1966) Some effects of artificial climates on the growth and chemical composition of plants. Am J Bot. 17: 416-482.

Ayenan MAT, Danquah A, Hanson P, Ampomah-Dwamena C, Sodedji FAK, Asante IK, Danquah EY (2019) Accelerating breeding for heat tolerance in tomato: An integrated approach. Agronomy, 9(11): 720.

Blakeslee AF, Belling J, Farhham ME, Bergner AD (1922) A haploid mutant in the Jimson weed *Datura stramonium*. Science 55: 646-47

Bortesi L, Zhu C, Zischewski J, Perez L, Bassié L, Nadi R, Forni G, Lade SB, Soto E, Jin X *et al.* (2016). Patterns of CRISPR/Cas9 activity in plants, animals and microbes. Plant Biotechnol J 14: 2203–2216.

Bugbee B, Koerner G (1997) Yield comparisons and unique characteristics of the dwarf wheat cultivar 'USU-Apogee'. Adv Space Res 20: 1891–1894.

Chari R and Church GM (2017) Beyond editing to writing large genomes. Nat Rev Genet 18: 749–760.

Chiurugwi T, Kemp S, Powell W, Hickey LT (2019) Speed breeding orphan crops. Theoretical and Applied Genetics 132(3): 607-616.

Croser JS, Pazos-Navarro M, Bennett RG, Tschirren S, Edwards K, Erskine W, Creasy R, Ribalta FM (2016) Time to flowering of temperate pulses *in vivo* and generation turnover *in vivo–in vitro* of narrow-leaf lupin accelerated by low red to far-red ratio and high intensity in the far-red region. Plant Cell Tissue Organ Cult127: 591–599.

Crossa J, Pérez-Rodríguez P, Cuevas J, Montesinos-López O, Jarquín D, de los Campos G, Dreisigacker S (2017) Genomic selection in plant breeding: methods, models, and perspectives. Trends in Plant Science 22(11): 961-975.

Dey SS, Bhatia R (2019) Speed breeding in vegetable crops, lecture delivered at winter school at ICAR-IARI New Delhi.

Doudna JAC, Charpentier E (2014) Genome editing. The new frontier of genome engineering with CRISPR-Cas9 28: 346.

Dwivedi SL, Britt AB, Tripathi L, Sharma S, Upadhyaya HD, Ortiz R (2015). Haploids: constraints and opportunities in plant breeding. *Biotechnology advances*, 33(6): 812-829.

Ghosh S, Watson A (2018) Speed breeding in growth chambers and glasshouses for crop breeding and model plant research. Nature protocol, https: //doi. org/10.1038/s41596-018-0072.

Guha S and Maheshwari SC (1964) *In vitro* production of embryos from anthers of Datura. Nature 212: 97-98.

Guha S, Maheshwari SC (1966) Cell division and differentiation of embryos in the pollen grains of datura in vitro. Nature 204: 497.

Hickey LT, Hafeez AN, Robinson H, Jackson SA, Leal-Bertioli SC, Tester M, Gao C, Godwin ID, Hayes BJ, Wulff BB (2019) Breeding crops to feed 10 billion. Nature Biotechnology. 37(7): 744-754.

Huihui Li, Awais Rasheed, Lee T. Hickey, and Zhonghu H. (2018). Fast-forwarding genetic gain, Trends in Plant Science 23: 3.

Liang Z, Chen K, Li T, Zhang Y, Wang Y, Zhao Q, Gao C (2017) Efficient DNA-free genome editing of bread wheat using CRISPR/Cas9 ribonucleoprotein complexes. Nature Communications (1): 1-5.

Mobini SH, Lulsdorf M, Warkentin TD, Vandenberg A (2015) Plant growth regulators improve in vitro flowering and rapid generation advancement in lentil and faba bean. In Vitro Cellular and Development Biology -Plant 51: 71-79.

Ray DK, Ramankutty N, Mueller ND, West PC, Foley JA (2012) Recent patterns of crop yield growth and stagnation. Nat Commun 3(1), 1-7.

Samineni S, Sen M, Sajja SB, Gaur PM (2020) Rapid generation advance (RGA) in chickpea to produce up to seven generations per year and enable speed breeding. The Crop Journal 8(1): 164-169.

Stetter MG, Zeitler L, Steinhaus A, Kroener K, Biljecki M, Schmid KJ (2016). Crossing methods and cultivation conditions for rapid production of segregating populations in three grain amaranth species. Frontiers in Plant Science 7: 816.

Sysoeva MI, Markovskaya EF, Shibaeva TG (2010) Plants under continuous light: a review. Plant Stress 4: 5-17.

Tanaka J, Hayashi T, Iwata H (2016) A practical, rapid generation -advancement system for rice breeding using simplified biotron breeding system; Breeding Science 66: 542-551.

Thomas WTB, Forster BP, Gertsson B (2003) Doubled haploids in Breeding. In: Doubled haploid production in crop plants. Springer. pp 337-349.

Voss-Fels KP, Stahl A, Hickey LT (2019) Modern crop breeding for future food security. BMC biology 17(1): 1-7.

Watson A, Ghosh S, Williams MJ, Cuddy WS, Simmonds J, Rey MD (2018) Speed breeding is a powerful tool to accelerate crop research and breeding. Nature Plants. doi: 10.1038/s41477-017-0083-8.

Weller JL, Beauchamp N, Kerckhoffs LHJ, Platten JD, Reid JB (2001) Interaction of phytochromes A and B in the control of de-etiolation and flowering in pea. Plant J 26: 283–294.

Yang F, Li Y (2017) The new generation tool for CRISPR genome editing: CRISPR/Cpf1. Sheng wu gong cheng xue bao Chinese J Biotechnol 33(3): 361.

Zhang Y. Liang Z, Zong Y, Wang Y, Liu J, Chen K, Qiu JL, Gao C (2016) Efficient and transgene-free genome editing in wheat through transient expression of CRISPR/Cas9 DNA or RNA. Nat Commun DOI: 10.1038/ncomms12617.

Chapter 7

Parthenocarpic Vegetables Importance and Approaches

B. Arunkumar, G.C. Nagesh, D.C. Manjunathagowda and R. Selvakumar

Introduction

Parthenocarpy (*Parthenos*, virgin; *karpos*, fruit) is the natural or artificial induction of fruit development without pollination and fertilization. For various parthenocarpic plant varieties, an increased supply of phytohormones to the gynoecium from sources other than the developing seeds has been reported to be sufficient to induce fruit growth (Abad and Monteiro, 1989; Garcia-Martinez, 1997; Gillaspy *et al.*, 1993). This observation suggests that parthenocarpic gene/s might primarily affect the hormone production, transport, and/or metabolism leading to sensitivity or higher hormone levels in the ovary capable of promoting fruit growth even in the absence of pollination and fertilization (Vivian-Smith *et al.*, 2001).

Parthenocarpy, like stenospermocarpy and apomixis are genetically controlled traits, in which fruit set is uncoupled from the series of events that occurs after pollination and fertilization. In stenospermocarpy, though distinct from parthenocarpy, normal pollination and fertilization do occur and initiate fruit development, but early embryo abortion results in seedless fruits as recorded in grapes and watermelon (Ledbetter and Ramming, 1989). Apomixis is embryo and seed development without fertilization and as a result apomictic species also produce fruit in the absence of fertilization as recorded in citrus, blackberry, mangoes and walnut (Koltunow and Grossniklaus, 2003).

Fruit development comprises early development and maturation (Figure 7.1). In most plants, normal early fruit development involves three phases 1: (1) fruit setting, (2) cell division, and (3) cell expansion. During the first phase, the ovary takes

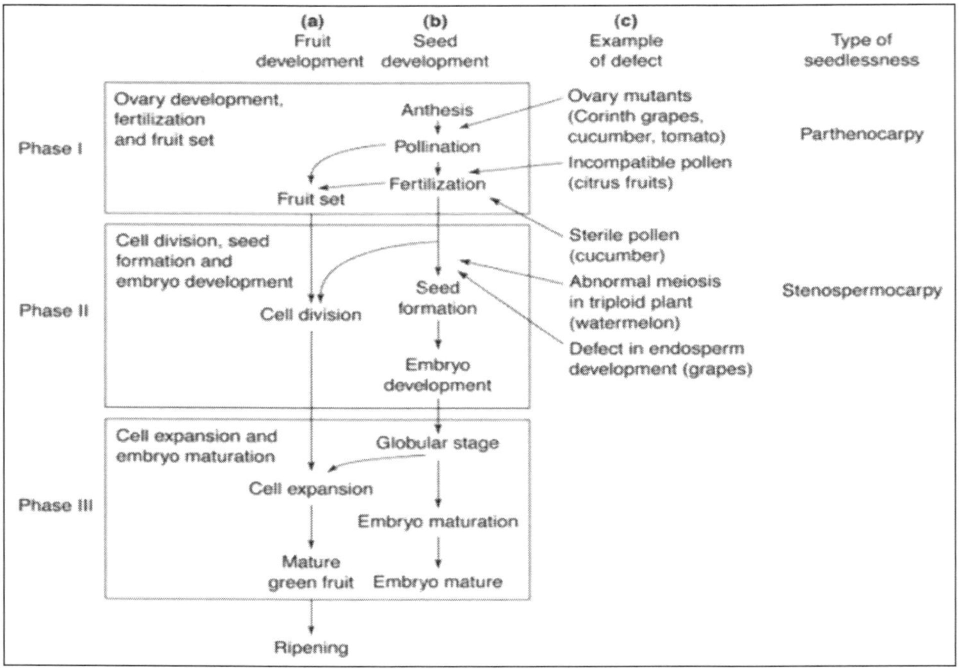

Figure 7.1. Schematic Representation of the Links Existing between Fruit and Seed Development.

(a) Different steps in fruit development, (b) Different steps in seed development, and (c) Examples of seedless fruit. Arrows drawn between (a) and (b) indicate the positive effect on fruit development of events preceding, or linked to, seed development. Arrows between (b) and (c) indicate the points in seed formation that might be deficient in seedless varieties.

the decision to either abort or to go further in fruit development. The next phase is the growth of the fruit as a result of cell division; during this phase, the increase in fruit size is low because the dividing cells are small and tightly compressed. The final size of a fruit will be highly dependent on the number of cells. The third and last phase begins after cell division ceases, the fruit grows by the increase in cell volume, until it reaches its final size. Cell expansion commonly increases fruit size by 100-fold, and this makes the greatest contribution to the final size of the fruit. At the end of early development, a green fruit is obtained, which has the size of a mature fruit, and the maturation phase occurs from this point onwards. During early fruit development, many pathways of communication between the sporophyte and the gametophyte are established. It is generally considered, for instance, that the decision of whether or not to set fruit is dependent on the successful completion of pollination and fertilization. The pollen produces gibberellins and it is well known that the application of exogenous gibberellins can induce an increase in the content of auxin in the ovary of an unpollinated flower of the tomato plant, and therefore trigger fruit setting in the absence of fertilization. Additionally, the developing embryo controls the rate of cell division in the surrounding fruit tissue, and there is evidence that the number of developing seeds influences the final size and weight of a fruit. It is generally considered that developing seeds promote cell expansion within the fruit by the production of auxin and other unknown molecules. Therefore, a developing seed has a very important role to play in the early development of fruit; producing seedless fruits might therefore be complicated.

Types of Parthenocarpy

Parthenocarpy can have a genetic basis (genetic or natural parthenocarpy) or it can be artificially induced.

Genetic Parthenocarpy

It is called obligatory when the expression of the parthenocarpy trait is not influenced by external factors. Genotypes with obligate parthenocarpy are unable to produce viable seeds either in the presence or in the absence of fertile pollen as reported in satsuma mandarin (George *et al.*, 1984). Plants with this condition are thought to be defective during female gametophyte development and require manual vegetative propagation. In contrast, facultative parthenocarpic plants produce seedless fruits only when pollination is prevented by growing conditions (*e.g.* temperature), or by manual manipulations (*e.g.* emasculation) or physiological barriers (*e.g.* self-incompatibility) (Ho and Hewitt, 1986). Various parthenocarpic mutants of tomato (*pat* 2, *pat3/4*), eggplants, squash, and pineapple are examples of facultative parthenocarpy (George *etal.*, 1984).

Artificial Induced Parthenocarpy

The induction of parthenocarpy is a rather common agricultural practice for some horticultural species (Schwabe and Mills, 1981). The use of irradiated pollen, natural or synthetic auxins and gibberellins has been rationalized as an exogenous (hormonal) signal able to increase in IAA content during ovary development (Tsao, 1980). In eggplant, the first increase takes place during the first five days after anthesis, while a major peak of IAA appears at 20 days after anthesis in both pollinated and auxin treated flowers (Lee *et al.*, 1997). These and similar data obtained with other species indicate that the exogenous signal (pollination/fertilization, phytohormones, auxin, analogues, *etc.*) triggers an increase in endogenous IAA content and activity in the ovary, which is responsible for promoting fruit set and development. It has been reported that parthenocarpic *pat*-2 tomato fruits show a slightly higher IAA content before anthesis, but more than 10 times higher level of IAA at anthesis in comparison to control fruits.

Importance of Parthenocarpy

- ☆ The independence of fruit development from fertilization is advantageous in horticulture when rate of fruit set is low. Pollen maturation and fertilization are affected by environmental factors such as light, temperature, relative humidity. Unfavourable environmental conditions can drastically depress fertilization and consequently fruit development. Parthenocarpic plants can circumvent this problem allowing seedless fruit development under environmental conditions adverse for pollination and/or fertilization. In horticulture, parthenocarpy can be exploited for increasing winter and early production; this means the possibility for the consumers to find fresh horticultural products in all seasons.

- Seedlessness is appreciated by consumers both in fruits for fresh consumption (*e.g.*, grape, citrus, and banana) as well as in conserved or processed fruits (*e.g.*, frozen eggplant, tomato sauce)
- Seedlessness can contribute to increase in the quality of fruits when seeds are hard or have a bad taste. In case of eggplant, the absence of seeds prevents browning and texture reduction of the pulp. Furthermore, seeds can produce substances that accelerate the deterioration of the fruit, as in watermelon and eggplant. In this regard, the absence of seeds can increase the shelf life of the fruits, allowing a better conservation. Parthenocarpy may also improve fruit quality through increases in the solid soluble content of the fruit.
- In the tomato sauce industry, seed content is a problem and seeds are removed to obtain sauce of good quality. As far as fruit quality concerns, productivity may also be increased in some seasons because fruit set and fruit growth are less affected under environmental conditions adverse for pollination and/fertilization including heavy rain, high humidity, hot and dry wind *etc.*
- Treating flowers with phytohormones or employing manual pollination is necessary for stable fruit setting in protected vegetable cultivation. However, these means are costly and/or labour-intensive. Use of parthenocarpic cultivar is the most cost-effective solution for stable fruit setting under sub-optimal environmental conditions.

Mechanism of Parthenocarpy

The process of fertilization induces an auxin-like signal in the pistil (O'Neill and Nadeau, 1997) which is dominant over the action of GA and results in a higher cellular expansion and an alteration of the mesocarp structure (Vivian-Smith and Koltunow, 1999; Figure 7.2). As pollination and seed formation via stimulation of the synthesis of active GA trigger fruit initiation and development, the artificial addition of GA should also initiate fruit set. This has been shown in a wide range of agricultural plant species (Schwabe and Mills, 1981).

The conditions adverse to pollination repress the processes of pollination and fertilization. Thus, the normal process of regulation of fruit and seed development does not occur. *pat-1*, *pat-2* and *pat-3/pat-4* genes responsible for parthenocarpy will directly stimulate one or more steps in the GA biosynthesis pathway, which results in enhanced expression of active GAs, and these GAs induce parthenocarpic fruit development (Figure 7.3).

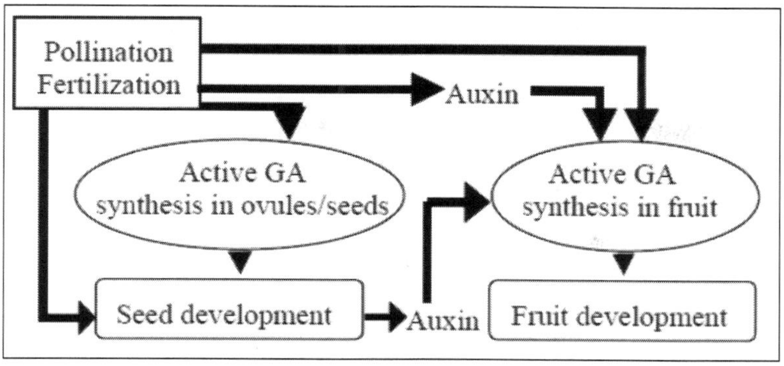

Figure 7.2. Model Proposed for the Regulation of Fruit and Seed Development.

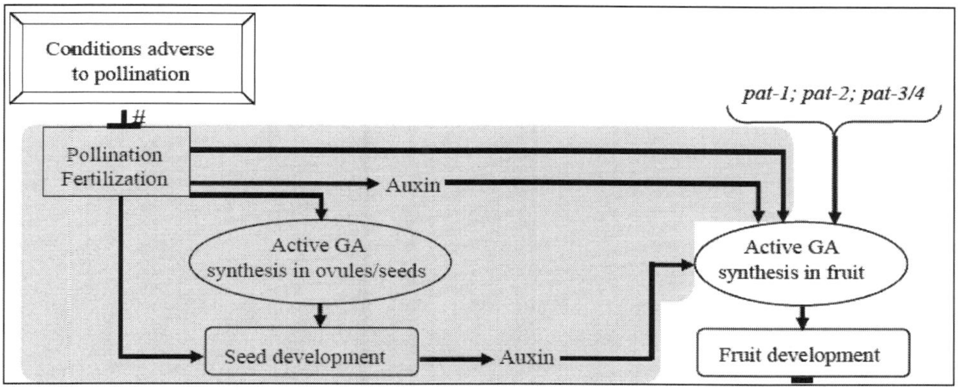

Figure 7.3. Model Proposed for the Parthenocarpic Tomato Fruit Development.

Approaches for Developing Parthenocarpic Vegetables

There are various approaches to obtain parthenocarpy in diverse plant species. Genetic occurrence of parthenocarpy can be exploited by using classical, molecular or mutation breeding approaches. Though parthenocarpy has a genetic basis, it can also be induced artificially in non-parthenocarpic plants by external application of hormones or by altering the endogenous plant hormone within ovules or carpel tissues.

- ☆ **Breeding approaches**
  - ❒ **Conventional**: Inter specific hybridization
    Intra specific hybridization
  - ❒ **Mutation**
  - ❒ **Molecular**
  - ❒ **Polyploidy**

} GENETIC INDUCTION

☆ **External application of PGR** } ARTIFICIAL INDUCTION
☆ **Transgenic approach**

Breeding Approaches

Classical or conventional breeding has been successfully applied to introduce parthenocarpy into economically important genotypes. Classical breeding comprises two fundamental steps; i) generating a breeding population that is segregating for the parthenocarpy trait of one parental genotype, and ii) selecting individual progeny from the segregating population that combine parthenocarpy with the desirable traits of the non-parthenocarpic parent. A breeding population is generated by using two types of crosses: inter- and intra-specific hybridization. Interspecific hybridization comprises variability in their offspring and is usually accompanied with sterility or semi sterility which helps in identifying genotypes that arise with strong parthenocarpic attributes or alleles (Gorguet *et al.*, 2008). Various parthenocarpic lines of tomatoes have been developed by using interspecific hybridization such as obligate parthenocarpy in aneuploid tomato developed from a cross between *Solanum esculentum* and *S. peruvianum* (Lesley and Lesley, 1941), IVT-line 1 was developed from a cross between *S. habrochaites and S. lycopersicum* (Zijlstra, 1985). Altered ploidy through interspecific hybridization is a common approach to obtain parthenocarpic fruits in various crops such as banana, watermelon and citrus (Fortescue and Turner, 2005). Usually, triploid plants cannot pass through meiosis, and are therefore sterile. Although the mechanisms of fruit initiation in triploid plants are largely unknown, the observations indicate that sterility is an important condition by which parthenocarpy becomes expressed. Parthenocarpic fruit set in various species associated with ploidy level. Ploidy alteration by interspecific hybridization results in parthenocarpic fruit set. The use of interspecific hybridization to generate *de novo* parthenocarpy is not successful in most of the *Solanaceae* species *e.g.* tomato, potato and eggplant (Carputo *et al.*, 2000). Crossing between diploid plant and its colchicine-derived tetraploid prevents triploid progeny to be produced due to the imbalance of parental ploidy. This condition is known as a "triploid block" which results in no fruits or some fruits with sterile seeds which is due to the failure of normal endosperm development (Contolini and Hughes, 1989). Intraspecific hybridization is commonly used to generate new lines and cultivars because they integrate high variability in their offspring, and this can form the basis of future selections (Lifschitz *et al.*, 1993). The use of intraspecific hybridization for producing a facultative parthenocarpic line suitable for a hot and dry climate (normal fruit at moderate temperature) was first introduced in tomato (Hawthorn, 1937). After that, various other parthenocarpic lines have been generated by using intraspecific hybridization *e.g.* Severenien, Oregon T5-4, Oregon Cherry, Oregon 11, Line 75/79, Line P-26, Line P-31, Line RG and IVT-line 2 in tomato (Baggett and Frazier, 1978).

Classical breeding is relatively inexpensive and technically simple. However, there are some limitations in the amount of genetic variation available within the crop (the gene-pool), and the approach requires large populations and multiple generations of selection to identify parthenocarpic plants that combine the best qualities of both parents. Moreover, with each cross, usually a large number of genes is transferred that may not be useful along with the desired ones in the target species (Beaver and Osorno, 2009). For minimizing these limitations as well as for the technical advancement, molecular and mutation breeding approaches were later introduced into the classical breeding.

Molecular Breeding

Molecular breeding has been introduced not so long ago into classical breeding methodologies. The purpose is to maximize the probability of successful allele retention by reducing the need of a large breeding population and multiple generations of selection (Huang *et al.*, 2002). Through Marker Assistance Selection (MAS), DNA markers detect the presence of allelic variation in genes underlying acertain trait and thus increase the efficiency and precision in the selection process. For example, many lines are discarded after MAS early in the breeding program, and this reduces the total numbers of lines to be screened and thus minimizes space and time constraints. Replacement of phenotype by markers allows off-season and heterozygote selection possible and thus makes the overall process more cost-effective by growing many generations per year (Ribaut and Hoisington, 1998). Integration of MAS with Quantitative Trait Loci (QTL) analysis has provided more potential for increasing breeding efficiency (Doerge, 2002). So far, only one parthenocarpy gene, *pat*, was mapped in tomato (Beraldi *et al.*, 2004). Recently, four QTLs associated with parthenocarpy were identified and mapped in tomato (Gorguet *et al.*, 2008). The isolation of these QTLs will enhance notonly our understanding about fruit set in tomato but also open possibilities to develop seedless fruits in other economically important species like *C. annuum* because of the high level of similarities within the *Solanaceae*.

Mutation Breeding

A major constraint in classical breeding is the lack of natural genetic variation in the genepool, which can be overcome by mutation breeding. A mutation is defined as a heritable change in sporophytic (*i.e.* in the somatic cells) or gametophytic tissue (*i.e.* germ cell line from where gametes arise) that is not caused by normal genetic recombination or segregation. Induction and selection of specific mutations can now be achieved (Cooper *et al.*, 2008), but spontaneous mutations have historically been an important source of genetic variation. Spontaneous mutations occur naturally and are used in classical breeding approaches. Good examples of this are the parthenocarpic *sha-pat* mutants in the tomato line 'Montfavet 191' (Pecaut and Philouze, 1978). However, when the genetic variation is limited, mutations have been induced chemically, by radiation treatment, or by DNA insertion (transposon, T-DNA). Various radiation treatments, such as helium accelerated ions in tomato (Masuda *et al.*, 2004), soft-X-ray in watermelon (Sugiyama and Morishita, 2000), gamma and thermal neutron irradiation in grapefruit (Vanamala *et al.*, 2007) and gamma irradiation in *Citrullus lanatus* (Sugiyama and Morishita, 2001) and *Citrus* (Spiegel- Roy and Vardi, 1989), have been used successfully to generate parthenocarpic mutants. For chemical mutagenesis two major classes of compounds are available: alkylating agents, such as ethyl methane sulphonate (EMS), and ethyl ethane sulphonate (EES), and base analogs, such as 5-bromouracil and 2-aminopurine. EMS is the most common mutagen, as it is very effective and convenient. It has been used to generate parthenocarpic mutants of Arabidopsis (*fwf*) and tomato (stock 2524: short anther mutant, *sha*) (Soressi, 1970; Motto *et al.*, 1975). Transposon mutagenesis has generally been a cis-genesis approach where mutagenesis in the recipient plant was induced by crossing with a transposon containing donor plant. The parthenocarpic mutants of apple have been identified where an already active and endogenous retro-transposon inserted into the intron region (intron 4 in Rae lime and intron 6 in Spencer seedless and Wellington Bloomless) and resulted in loss of function of MADS-box transcription factor gene *MdPI* (Yao *et al.*, 2001). Although mutation breeding is effective to obtain parthenocarpic fruits, its applicability has been limited in the condition where parthenocarpy is associated with pleiotropic or unfavorable genetic link characteristics (Huang *et al.*, 2002; Pandolfini *et al.*, 2002; Yao *et al.*, 2001).

Artificial Induction

Artificial induction of parthenocarpic fruit set is useful to understand the role of plant hormones in fruit set and fruit development. Parthenocarpy can be induced artificially in non-parthenocarpic plants by external application of plant growth regulators or by altering the endogenous plant hormone levels within ovules or carpel tissues, as has been shown in arabidopsis, tomato, eggplant, watermelon, and citrus (Rotino *et al.*, 1997; Carmi *et al.*, 2003; Mezzetti *et al.*, 2004).

External Application of Plant Growth Regulators (PGRs)

A very effective way to induce parthenocarpy has been reported by exogenous application of PGRs to the stigma or the carpels. Auxins (IAA; or it synthetic analogs 2, 4-D and 4- chloro-IAA, NAA), gibberellins (primarily GA1, GA3, GA4 and GA20) and to a lesser extend cytokinins are known to induce parthenocarpy in several crops (Kojima *et al.*, 2003).

Transgenic Approach (Genetic Engineering of Parthenocarpy)

Spena and Rotino (2001) reviewed the potential applications of recombinant DNA technology for the development of parthenocarpic plants. They made a distinction between two approaches: the first is based on unbalanced embryo development and/or prevention of seed production, and the second consists of modulating the phytohormone content and/or activity in a specific organ in order to trigger parthenocarpic fruit development. As stated earlier, parthenocarpic fruit development is mainly dependent on the level of auxin and gibberellin in the ovary during anthesis and fruit development. This has been demonstrated with parthenocarpic mutants and with parthenocarpic fruit development after exogenous application of auxin or gibberellin. It is therefore expected that parthenocarpy can be induced in transgenic plants in which a gene involved in the biosynthesis of either of these two plant growth hormones is expressed in the ovary (Varoquaux *et al.*, 2000). Such transgenic plants should contain a gene which gives an increase in the level of a specific hormone and is under the control of an ovary-specific promoter that is active during appropriate stages of flower development.

Genetic engineering aimed at obtaining parthenocarpic fruits was first successful in tobacco and eggplant, with the introduction of the *iaaM* gene, a auxin synthesising gene from *Pseudomonas syringae* pv. *savastanoi*. This gene is under control of the placenta and ovule specific *DefH9* promoter from *Anthirrhinum majus* (Rotino *et al.*, 1997). The *iaaM* gene codes for a tryptophan mono-oxigenase that converts the tryptophan into indolacetamide, which is then converted to auxin. The *DefH9* promoter is active during early stages of flower development. Seedless fruits were produced from emasculated flowers, and seeded fruits from pollinated flowers.

Tomato

Tomato is one of the most important vegetable crop of the world. Tomato fruits are consumed either fresh or processed. During winter cultivation in greenhouse the flowers are treated with auxinic phytohormones which cause parthenocarpic fruit development and production under environmental conditions (10-15°C short day and low light intensity) adverse for fruit set and growth (Schwabe and Mills, 1981). However, either higher sensitivity to auxins or an excess of exogenous phytohormones causes malformations of the tomato fruit (Santangelo and Soressi, 1990). Consequently, breeding programs for fresh market tomatoes have usually screened tomato lines for an optimal response of the flowers to phytohormonal sprays.

The Three Main Sources of Parthenocarpy in Tomato

Detailed reviews on the different types and sources of parthenocarpy in tomato were presented by George *et al.* (1984) and Lukyanenko (1991). Three sources have been widely studied because of their perspectives for practical application to produce seedless fruits: *'Soressi'* or *'Montfavet 191'*, *'Severianin'* and *'RP75/59'*. Philouze (1985) considered these three sources as the only ones able to give parthenocarpic fruits after emasculation, with nearly the same properties as fruits obtained after pollination and fertilization.

The *pat-1* Gene

Soressi and Salamini (1975) described the short anther (*sha*) mutant in tomato. This mutant has abnormal stamens and produces parthenocarpic fruits. This parthenocarpic phenotype was thought to be caused by two closely linked recessive genes, *sha* and *pat-1* (*parthenocarpic fruit-1*). However, Pecaut and Philouze (1978) independently obtained a mutant with the same phenotype, designated *'Montfavet 191'*. These independent mutations were allelic (Pecaut and Philouze, 1978; Philouze and Pecaut, 1986) and the characters 'short anthers' and 'parthenocarpic fruits' could never be separated. It was concluded that the described phenotype was caused by a recessive mutation in a single gene, designated *pat-1*, with pleiotropic effects.

Studies have shown that the start of ovary growth in *pat-1* mutants occurs at the pre-anthesis floral stage compared to two days post anthesis in the wild type (Mazzucato *et al.*, 1998). This results in a higher ovary weight and higher numbers of pericarp cell layers in *pat-1* ovaries (Mapelli *et al.*, 1978). This *pat-1* mutant also shows irregular meiosis which results in a lower number of viable female gametes (Mazzucato *et al.*, 1998). Apparently, aberrations in the ovule production play a role in seed set of *pat-1* fruits. However, fertilization in *pat-1* ovaries is strongly impaired, even in the ovules that appear normal, probably due to a defective pollen tube-placenta interaction (Mazzucato *et al.*, 2003). Therefore, even in conditions favorable for seed production, *pat-1* genotypes give a very low seed set (Mazzucato *et al.*, 1998; Philouze and Pecaut, 1986). The *pat-1* mutant produces tomato fruits which are about two-thirds the normal size and weight (Bianchi and Soressi, 1969; Falavigna *et al.*, 1978; Philouze and Pecaut, 1986). A reduction in fruit size combined with difficulties to obtain seeds has made the *pat-1* gene less attractive for breeding (Philouze and Pecaut, 1986). Recently, the *pat-1* gene has been mapped on the long arm of chromosome 3 (Beraldi *et al.*, 2004), opening the way towards cloning of the gene.

The *pat-2* Gene

Another source of parthenocarpy has been found in the tomato cultivar '*Severianin*'. Philouze and Maisonneuve (1978) and Nuez *et al.* (1986) showed that a single recessive *pat-2* gene was responsible for parthenocarpy in '*Severianin*' while Vardy *et al.* (1989b) hypothesized a model based on two recessive genes, one having a major effect (*pat-2*) and one a minor effect (*mp*). Philouze *et al.* (1988) showed that the choice of the recurrent parent in which the *pat-2* gene is introduced is very important for plant vigor. In the genetic background of the tomato line *Apedice*, the plants are smaller and less vigorous, resulting in a reduction in yield, fruit set, and firmness of the fruits (Philouze *et al.*, 1988). No differences in yield and vigour were found when *pat-2* was introduced into the genetic background of tomato lines '*Monalbo*' or '*Porphyre*'. Parthenocarpy may interact with growth habit. Lin *et al.* (1984) found that in a population segregating for growth habit and *pat-2*, the majority of the parthenocarpic plants had determinate growth. However, no genetic linkage was found between parthenocarpy and the locus controlling the determinate growth habit (*sp*). Probably, determinate growth allows better expression of parthenocarpy. The main advantage of exploiting parthenocarpy from *Severianin* is the potential that the deleterious pleiotropic effects of *pat-2* can be overcome by a suitable genetic background.

The *pat-3* and *pat-4* Genes

The tomato line *RP75/59* was found in the progeny from a cross between 'Atom' x 'Bubjekosoko'. A high level of parthenocarpy was observed without obvious effects on fruit size and appearance. However, in this new source of parthenocarpy an association was observed between seed set and size of fruits within the same truss or plant: When both seeded and seedless fruits are present on the same plant, the seeded fruits grow bigger than the seedless fruits. Genetic studies on *RP75/59* showed that two to five genes control this trait. (Nuez*et al.*, 1986; Vardy *et al.*, 1989). The differences in the number of genes involved in parthenocarpy are probably due to the very difficult assessment of the trait. *RP75/59* is facultative parthenocarpic, therefore, to obtain seedless fruits, the plants have to be grown in conditions adverse to pollination and/or fertilization, or the flowers have to be emasculated. Nowadays, a genetic model with the two genes *pat-3* and *pat-4* in *RP75/59* is commonly accepted. Philouze (1989) showed that *pat-2* and *pat-3/pat-4* are not allelic. The polygenic inheritance of the RP75/59 parthenocarpy source and the different fruit sizes of seeded and seedless fruits on the same truss make the use of *pat-3/pat-4* genes less attractive for breeding.

Table 7.1. Potentially Useful Sources of Parthenocarpy in Tomato

Cultivar/Breeding Line	*Country of Origin*	*Genetic Control*	*Reference*
PI190256	New Caledonia	Unknown	Johnson and Hall (1954)
Sub arctic Plenty	Canada	Unknown	Charles and Harris (1972)
Oregon T5-4	USA	Unknown	Baggett and Frazier (1978)
Oregon Cherry	USA	Unknown	Baggett and Frazier (1978)
Oregon 11	USA	Unknown	Baggett and Frazier (1978v
Stock 2524	Italy	*pat*	Soressi (1970)
Montafavet 191	France	*pat*	Pecaut and Philouze (1978)
Severianin	Russia	*pat-2*	Philouze and Maisonneuve (1978)
RP75/59	West Germany	Unknown[1]	Philouze and Maisonneuve (1978)
PSET-1[2]	USA	*pat-2*	Scott and George (1983)

Mapping and Characterization of Novel Parthenocarpy QTLs in Tomato

Gorguet *et al.* (2008) studied the genetics of parthenocarpy in two different lines, IL5-1 and IVT-line 1, both carrying *Solanum habrochaites* chromosome segments. Parthenocarpy in IL5-1 is under the control of two QTLs, one on chromosome 4 (*pat4.1*) and one on chromosome 5 (*pat5.1*). IVT-line 1 also contains two parthenocarpy QTLs, one on chromosome 4 (*pat4.2*) and one on chromosome 9 (*pat9.1*). In addition, they also identified one stigma exsertion locus in IL5-1, located on the long arm of chromosome 5 (*se5.1*). It is likely that pat4.1, from IL5-1 and pat4.2, from IVT-line 1, both located near the centromere of chromosome 4 are allelic. By making use of the microsynteny between tomato and arabidopsis in this genetic region, the authors confirmed that ARF8 as a potential candidate gene for these two QTLs. *ARF8* is known to act as an inhibitor for further carpel development in *Arabidopsis*, in absence of pollination/fertilization. Expression of an aberrant form of the arabidopsis *ARF8* gene, in tomato, has been found to cause parthenocarpy. This candidate gene approach may lead to the first isolation of a parthenocarpy gene in tomato and will allow further use in several crop species.

Genetic Engineering of Parthenocarpic Fruit Development in Tomato

Parthenocarpy was engineered in two genotypes of *Lycopersicon esculentum* Mill. by using the *DefH9-iaaM* chimeric gene. The parthenocarpic trait consists of fruit set and growth in the absence of fertilization. Seedless parthenocarpic fruits were obtained from emasculated flowers, and fruits with seeds from pollinated flowers (Ficcadenti *et al.*, 1999).

Cucumber

Parthenocarpy is an economically important yield- and quality-related trait in cucumber. Parthenocarpy is regulated by endogenous plant growth regulators (*e.g.*, diffusible auxin, IAA), and their balance is dramatically influenced by environment (Kim *et al.*, 1994). Phenotypic selection has, however, resulted in the development of parthenocarpic hybrids (More and Budgujar, 2002) and genetic stocks. It is clear that parthenocarpy is genetically controlled, but there is little agreement regarding the number and type of gene action involved. Meshcherov and Juldasheva (1974) suggested that parthenocarpy is recessive and controlled by a single gene. Kvasnikov *et al.* (1970), however, proposed that many incompletely recessive genes control parthenocarpy. Pike and Peterson (1969) simultaneously proposed that a single dominate gene expressing incomplete dominance controls parthenocarpy in cucumber. Results of de Ponti and Garretsen (1976) indicate that parthenocarpy may be quantitatively inherited. In fact, studies by Sun *et al.* (2006) indicate that the genetics of parthenocarpy are complex. Generation means analyses in cross-progeny derived from elite processing cucumber lines indicated gene action generally could not be adequately explained by a simple additive-dominance model. Moreover, the analysis of F_3 families indicated that more than five genes control parthenocarpy, and that growing environment and epistatic interactions dramatically influence trait expression.

Three tropical gynoecious cucumber lines were developed first time in India at Indian Agricultural Research Institute, New Delhi by More and Seshadri (1988). Further efforts were diverted to develop the parthenocarpic tropical gynoecious cucumber lines with yellow skinned fruits similar to that of cv. Poona Khira primarily grown in Maharashtra state. Thirteen lines (F_6) with yellow skinned fruits and six lines which produce green skinned fruits and true breeding for gynoecy were isolated from a complex cross [(Table Green 68 x SC3, F_2) x Poona Khira] F_3 x [(GY 14 x Table Green 68, F_1) x Poona Khira] F_3.These 19 lines were further evaluated for skin colour, gynoecy and parthenocarpic nature during Kharif, 1996 and Spring-Summer, 1997.

Among 13, four yellow skinned parthenocarpic tropical gynoecious lines *viz.*, PKG-1-2, PKG-1-11, PKG-1-12 and PKG-1-15 isolated which showed 100 per cent parthenocarpy (Figure 7.4). The remaining yellow fruit skinned tropical gynoecious lines exhibited 87.50 to 96.43 per cent parthenocarpy. Three out of six green fruit skinned parthenocarpic tropical gynoecious lines namely, PKG-1-21, PKG-1-23 and PKG-1-24 exhibited 100 per cent parthenocarpy. Other three green fruited tropical gynoecious lines showed 82.69 to 94.74 per cent parthenocarpy. The highest per cent of fruit setting (80 per cent) was noticed with PKG-1-15. Fruit setting in rest of yellow fruited and green fruited parthenocarpic tropical gynoecious lines ranged from 38.2 to 75.0 and 44.2 to 64.0 per cent, respectively. Fruit length in yellow fruit skinned and green fruit skinned parthenocarpic tropical gynoecious lines ranged from 17.6 to 27.5 and 19.9 to 26.2 cm (More and Budgujar, 2002).

Mapping Parthenocarpy in Cucumber

An $F_{2:3}$ matingdesign was utilized to map quantitative trait loci (QTL) for parthenocarpy using a narrow cross employing two gynoecious, indeterminate and normal leaf lines [2A (parthenocarpic) and Gy8 (non-parthenocarpic)]. QTL

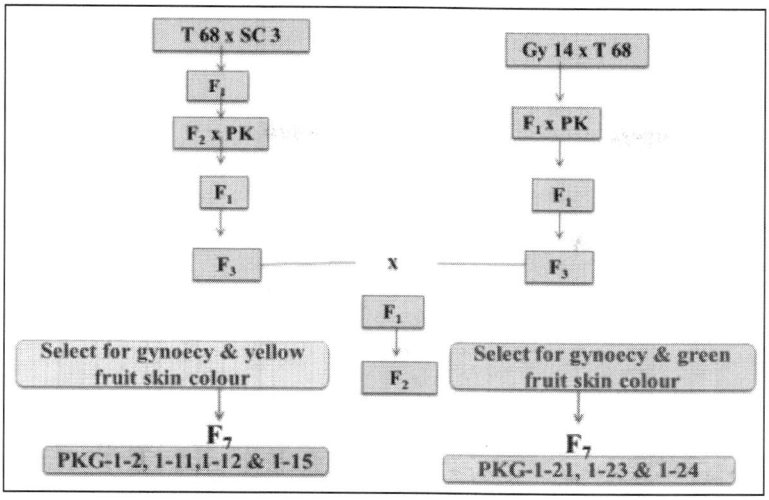

Figure 7.4. Development of Parthenocarpic Tropical Gynoecious Lines (More and Budgujar, 2002).

detection was performed employing 2A- and Gy8- coupling phase data using the parthenocarpic yield of 126 F_3 families grown at two locations at Hancock, WI in 2000. The QTLs detected in this study were compared with the map locations of QTLs conditioning first-harvest yield of seeded cucumber characterized in a previous study. There were 10 QTLs for parthenocarpy detected defining four genomic regions, in which three QTLs also mapped to the same genomic regions as QTLs detected for fruit yield at first harvest as reported in a previous study. The eight fluorescence amplified fragment length polymorphism (AFLP) markers linked to parthenocarpy through QTL mapping defined herein (four each in linkage groups 1 and 4) are candidates for use in marker-assisted selection programmes where breeding for increased levels of parthenocarpy is an objective in the elite-processing cucumber populations (Sun *et al.*, 2006).

Maintenance of Parthenocarpic Gynoecious Lines

The models for the regulation of sex expression of cucumber in the point of view to breeding and seed production is based on the interaction of hormonal and genetic system of the plants. Femaleness is promoted by ethylene, auxins, *etc.*, while maleness is induced by gibberellic acid and other chemicals counteracting ethylene. Sex expression and induction of male flowers depend on $AgNO_3$ concentration and the number of treatments. Elizabeta and Susaj (2010) studied the effect of different concentrations of $AgNO_3$ in inducing male flowers in parthenocarpicgynoecious lines and suggested that the highest number of staminate flowers was taken on plants which were treated twice or three times with 400-500 ppm $AgNO_3$ in seven days intervals. On the other hand, all treatments used once were ineffective for producing staminate flowers after tenth node. In some cases, some days after treatments with 400 and 500 ppm, were observed injuries which were recovered for seven to ten days.

Brinjal

In eggplant (*Solanum melongena* L.) cultivation, high temperatures promote growth of the plants, but very high temperatures are not suitable for the formation and growth of flower buds and are apt to induce pollen. Such damage to the flowers results in decreased pollination rates, reduced fruiting, and increased fruit abscission and subsequently, a reduced harvest and reduced fruit quality. To overcome such difficulties, plant hormones have often been used in eggplant cultivation under unfavorable conditions. Hormone treatment requires a great deal of labor (typically from one-fourth to one third of the total work), however, and is thus expensive. To mitigate these defects in fruit set under unfavourable conditions, the breeding of parthenocarpic cultivars may be a cost-effective alternative

Table 7.2. Genetics of Parthenocarpy in Brinjal

Genotype	*Gene Action*	*References*
Talina x EPL1	Single major gene	Yoshida *et al.* (1998)
Mileda x ASL-1	Single dominant gene	Kuno and Yabe (2005)

In an eggplant-breeding program in Japan, the parthenocarpic trait of 'Talina' was introduced into the Japanese cultivars 'Nakate Shinkuro' and 'Nasu Chuukanbohon Nou 1', which has *Fusarium* wilt resistance, and parthenocarpic breeding lines with favorable fruit quality have been developed. Crosses were conducted between 'Talina' (a commercial parthenocarpic F_1) and both 'Nakate Shinkuro' and 'Nasu Chuukanbohon Nou 1'.True-bred parthenocarpic lines (F_7 and later generations) were selected (denoted 'AE-P' lines). Pedigree of the lines is presented in Figure 7.5. 'AE-P01', 'AE-P02', and 'AE-P10' resulted from the crossing of 'Talina' with 'Nasu Chuukanbohon Nou 1';'AE-P03', 'AE-P05', and 'AE-P06' resulted from the crossing of 'Nakate Shinkuro' with 'Talina'; and 'AE-P08', 'AE-P11', 'AEP12', and 'AE-P14' resulted from the crossing of ('Nakate Shinkuro' x 'Talina') with ('Talina' x 'Nasu Chuukanbohon Nou 1'). We compared the fruiting abilities of the 10 AE-P lines with those of a parthenocarpic line ('Talina 2/1', a double-haploid line produced from 'Talina') (Kikuchi *et al.*, 2008).

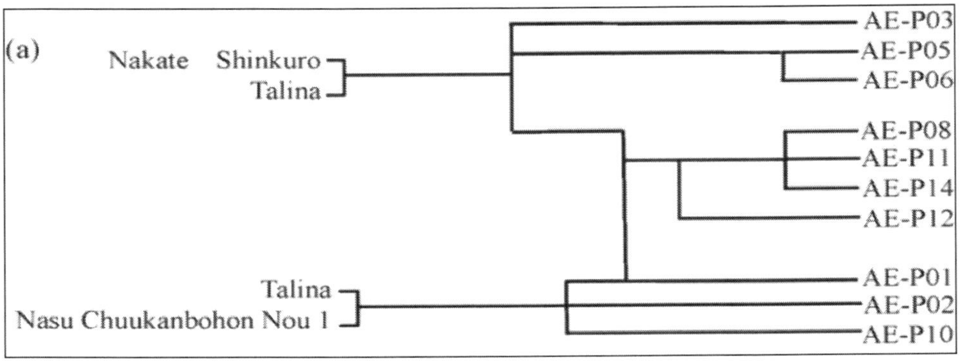

Figure 7.5. Pedigrees of the Parthenocarpic Eggplant Lines (Kikuchi *et al.*, 2008).

"Anominori", a F_1 hybrid between two parthenocarpic inbred lines, AE-P08 and AE-P01 was developed. In order to compare the parthenocarpic ability among three cultivars, the development of fruits after excision of styles was observed. Parthenocarpic fruits of marketable size and weight were produced from 55 to 60.7 and 58 per cent of flower buds after excision of styles. However, no fruits of marketable size set in European parthenocarpic cultivars, Mileda or Talina 2/1. No fruits of marketable size were also produced in non-parthenocarpic Senyou Nigou, Chikuyou or Nakate Shinkuro (Kikuchi *et al.*, 2008).

Genetic engineered parthenocarpic hybrids P1, P3 and P4 were obtained by crossing (as male parents) the primary transgenic plants DR2 *DefH9-iaaM* 34-1with the lines Tal 1/1 (P1) and Tina (P3) and Tal 1/1*DefH9-iaaM* 1-1 with the line Tina (P4). The transgenic hybrids were compared to the untransformed control hybrids C1 (Tal 1/1xDR2) and C2 (Tal 1/1xTina) and the commercial Talina F_1 hybrid. Therefore, the hybrids P1 and P4 have the identical genetic constitution as C1 and C2 but for the presence of the transgene. Winter production of three eggplant hybrids transgenic for the parthenocarpic gene *DefH9-iaaM* was compared, in an unheated greenhouse. Each hybrid was either treated or untreated with a commercial formulation of phytohormones to induce fruit set and growth. Early winter production consisted of the first four harvests, corresponding to the period with the environmental conditions most limiting for fruit set and growth. In this period, the average temperature ranged from 11 to 16 °C, the average maximum light intensity ranged from 310 to 360 W/m^2 and the average soil temperature declined from 19 to 17 °C, the lowest values of the whole growing season. In the plots not treated with hormones, the untransformed hybrids gave a very poor yield, which considerably increased when their flowers were sprayed with the commercial hormonal formulation. In contrast, the transgenic parthenocarpic hybrids were not affected by the hormonal sprays and gave a similar production whether or not their flowers had been sprayed. The untreated transgenic parthenocarpic hybrids P1 and P4 yielded a significant higher production (*i.e.* an increase of 13 and 8 tonnes/ha, respectively) in comparison to their correspondent sprayed controls C1 and C2. It is worthwhile to point out that the presence of the transgene is the only genetic difference between the P1 and P4 hybrids and the C1 and C2 controls, respectively. Moreover, the data show that the hormonal treatment also improved the productivity of the parthenocarpic cultivar Talina. During the subsequent ten harvests, performed in the remaining cultural cycle (4 March–1 June), the transgenic hybrids out yielded the untransformed controls, and the hormonal sprays had no effect on their productivity(Donzella, *et al.*, 2000).

Capsicum

Parthenocarpy is a desirable trait in *Capsicum annuum* production because it improves fruit quality and results in a more regular fruit set. Almost all seedless fruits obtained either by only emasculation or emasculation followed by hormone application showed stronger growth of carpel like structures (CLS) compared to seeded fruits. Structural analogy of CLS with *bel1* mutant of Arabidopsis suggests that CLS are transformed from abnormal ovules. *Capsicum* genotypes with high parthenocarpic potential showed a stronger CLS development suggesting a relation between female sterility and parthenocarpy. The parthenocarpic potential appeared to be controlled by a single recessive gene. The CLS phenotype and parthenocarpy could not be linked to a single locus, suggesting that absence of fertilization induces parthenocarpic fruit development and allows CLS growth, which substitutes developing seeds in promoting fruit development.

Tiwari *et al.* (2011) evaluated sweet pepper germplasm to identify parthenocarpic genotype(s) of sweet pepper with the ability to produce good quality fruits without seeds. Eleven genotypes of sweet pepper were studied at normal (20/20°C D/N) and at low night temperature (20/10°C D/N), in two greenhouse compartments. A higher percentage of parthenocarpic fruits was observed for all genotypes at low night temperature as compared to normal temperature. For Line 1 and Line 3, percentage of parthenocarpic fruit was high at both low and normal night temperature, while substantial parthenocarpic fruit growth was observed for Lamuyo A, Lamuyo B, Gen A, Gen B, Gen C and Bruinsma Wonder only at low night temperature.

Inheritance of Parthenocarpy in Capsicum

To study the genetic basis and inheritance of the parthenocarpic potential in *C. annuum*, the parthenocarpic genotype Line 3 was crossed with the non-parthenocarpic parents Lamuyo B, OR F_2 (a male sterile plant selected from an F_2 population) and Parco. Since Line 3 is a small fruited (average fruit weight of 121 g) and Lamuyo B is a large fruited genotype (average weight of 208 g) for seeded fruit, fruit size traits segregated independently upon crossing. In the F_2 analysis, a plant was considered parthenocarpic when emasculated flowers all produced seedless fruits showing a shiny appearance. In all three F_2 populations parthenocarpic plants were observed in 1:3 ratios. Furthermore, when the F_1 of Line 3 x Lamuyo B was backcrossed with Line 3, parthenocarpy was observed in a 1:1 ratio. These data support the hypothesis that parthenocarpy present in Line 3 is controlled by a single recessive gene (Tiwari *et al.*, 2011).

Watermelon

Seedless cultivars are produced by crossing a tetraploid (4X=44) inbred line as the female parent with a diploid (2X=22) inbred line as the male parent of the hybrid. The reciprocal cross (diploid female parent) does not produce seeds. The hybrid is a triploid (3X=33) and is female and male sterile. Triploid plants have three sets of chromosomes, and three sets cannot be divided evenly when they go into two daughter cells during meiosis (the cell division process that produces the gametes). Since the triploid hybrid is female sterile, the fruit are seedless. Because the triploid is also male sterile, it is necessary to plant a diploid cultivar in the production field to provide the pollen that stimulates fruit to form. Usually, one third of the plants in the field are diploid and two thirds are triploid. Cultivars should be chosen that can be distinguished easily so the seeded diploid fruit can be separated from the seedless triploid fruit for marketing.

Table 7.3. Triploid Watermelon Varieties Developed in India

Variety	*Released Institute*
Pusa Bedana	IARI, New Delhi
Arka Madhura	IIHR, Bangalore
Red seedless	KAU, Thrissur
Yellow seedless	(Yet to be released)

A new method for producing seedless watermelon in diploid plants using soft-X- irradiated pollen is described. Fruit set at almost the same rate despite soft-X-irradiation. Empty seeds were produced in the watermelon cultivars 'Benikodama' and 'Fujihikari TR' following hand pollination using soft-X-irradiated pollen. Soft-X- irradiation doses of 800-1000 Gy resulted in small empty seeds in 'Fujihikari TR', whereas 400-1000 Gy doses gave the best results for 'Benikodama' (Table 10) (Sugiyama and Marishitha, 2000).

Achievements

Table 7.4. Parthenocarpic Varieties of Tomato and Brinjal

Tomato	*Brinjal*
Severianin	AE-PO3
Oregon Pride	Talina
Oregon Sugar	Mileda
Montfavet	Anominori

Table 7.5. Parthenocarpic Cucumber Varieties/Lines from Public Sector

Sl.No.	*Name of the Variety/Line*	*Source*
1	Pant Parthenocarpic Cucumber-1	GBPUAT, Pant Nagar
2	Pant Parthenocarpic Cucumber-2	GBPUAT, Pant Nagar
3	Pant Parthenocarpic Cucumber-3	GBPUAT, Pant Nagar
4	DPaC-6	IARI, New Delhi
5	DPaC-9	IARI, New Delhi
6	DPaC-10	IARI, New Delhi

Table 7.6. Parthenocarpic Cucumber Hybrids from Private Sector

Sl.No	*Name of the Hybrid*	*Source of Seed*
1	Isatis	Nunhems India Pvt. Ltd
2	Hilton	Nickerson Zwaan
3	PY-1026	East West Seeds India Pvt. Ltd
4	NS-492	Namdhari Seeds India Pvt. Ltd

Sl.No	Name of the Hybrid	Source of Seed
5	NS-498	Namdhari Seeds India Pvt. Ltd
6	NS-499	Namdhari Seeds India Pvt. Ltd
7	Claudia	Tropic Seeds Pvt. Ltd
8	Aviva	Tropic Seeds Pvt. Ltd
9	Asma	Tropic Seeds Pvt. Ltd
10	Kian	Nunhems India Pvt. Ltd

Limitations

Two main problems have so far strongly limited the development of parthenocarpic vegetables.

1. **Stability and uniformity of parthenocarpic production are difficult to obtain in vegetable crops**. Several sources of parthenocarpy are known in tomato. However, they are often associated with unfavourable characteristics or are poorly expressed in indeterminately growing plants. In addition, when seeded and seedless fruits develop on the same tomato cluster, seeded fruits are generally larger. A logical approach to overcome this problem would be to combine several parthenocarpy genes to obtain a higher level of parthenocarpy, but pyramiding genes is impossible without molecular assisted selection.
2. **Parthenocarpy hampers the production of commercial seeds**. The development of tomato lines with a high parthenocarpy level is problematic for commercial seed production. With a high level of parthenocarpy, the swelling of the ovary, characteristic of fruit setting, often starts before anthesis. Therefore, at anthesis the ovary cannot be pollinated anymore and the production of seeds becomes impossible.

Conclusion

Pollination and fertilization are required in most flowering plants to initiate the transition from a fully receptive flower to undergo fruit development. After fertilization the ovules develop into seeds and the surrounding carpels develop into the fruit, while in the absence of fertilization the ovules degenerate and growth of the surrounding carpels remains repressed. The initiation of fruit set can be uncoupled from fertilization, and these results in the development of seedless or parthenocarpic fruits. This can be achieved by ectopic application or artificial overproduction of plant hormones, or by mutating or altering the expression of specific genes.

Inducing parthenocarpic fruits by transforming plants with a transgene composed of a fusion between an ovary-specific promoter and an auxin precursor biosynthesis gene has produced impressive results. Parthenocarpic eggplants with no deleterious pleiotropic effects have been effectively produced. This strategy seems to be able to induce parthenocarpy in a wide variety of species. This is surprising because, theoretically, parthenocarpy was the basis of the type of seedlessness that was most difficult to achieve mainly because fruit development must take place without fertilization and sometimes even with no stimulation via pollen deposition.

One question, which has not yet been fully answered, is how the induction of auxin in the ovary is able to substitute for seeds in the developing fruit. In tomato, several natural parthenocarpic cultivars exist, but all of them have disadvantages that are caused by the pleiotropy of the genes involved. In the future, because of the availability of transgene technology, it thus seems pointless to try to find and use new parthenocarpic mutants in crops.

In conclusion, the interest of consumers, farmers and seed companies in seedless fruit and the technical progress of genetic engineering of seedlessness lead us to believe that, in the near future, seedlessness could been improvement introduced into a wider range of fruits and vegetables.

Future Strategies

- Combining several parthenocarpy genes to obtain a higher level of parthenocarpy through gene pyramiding.
- Combining parthenocarpy with male sterility

A combination of parthenocarpy with pollen sterility has the advantage of preventing the presence of seeded fruits and therefore would lead to a more uniform production of seedless fruits. However, the seed production would become impossible. A way to solve this problem is by the use of functional sterility, which is characterized by a normal development of viable pollen, but natural pollination is strongly restricted due to some deviation from the normal morphology and function of the flower. One of the best functional male sterility genes in tomato is *positional sterility 2* (*ps-2*), which is characterized by non-dehiscent anther bags.

The combination of parthenocarpy with *positional sterility-2* would prevent the development of seeded fruits and the production of seeds would still be possible by manual anther opening. Breeding for parthenocarpy in a functionally sterile background would focus on the parthenocarpic potential of the plants and not on a selection against seed set as it has been done so far.

REFERENCES

Abad M, Monteiro AA (1989) The use of auxins for the production of greenhouse tomatoes in mild-winter conditions - a review. *Sci Hortic* 38: 167-192.

Baggett JR, Frazier WA (1978) Oregon T5-4 parthenocarpic tomato line. *HortScience* 13: 599

Beaver JS, Osorno JM (2009) Achievements and limitations of contemporary common bean breeding using conventional and molecular approaches. *Euphytica* 168: 145-175.

Beraldi D, Picarella ME, Soressi GP, Mazzucato A (2004) Fine mapping of the Parthenocarpic fruit (*pat*) mutation in tomato. *Theor Appl Genet* 108: 209-216.

Carmi N, Salts Y, Dedicova B, Shabtai S, Barg R (2003) Induction of parthenocarpy in tomato via specific expression of the *rolB* gene in the ovary. Planta 217: 726-735.

Carputo D, Barone A, Frusciante L (2000) 2n gametes in the potato: essential ingredients for breeding and germplasm transfer. *Theor Appl Gen* 101: 805-813.

Charles WB, Harris RE (1972) Tomato fruit set at high and low temperatures. *Can J Plant Sci* 52: 497-506.

Contolini CS, Hughes KW (1989) Reciprocal differences in intraspecific crosses of tobacco result from embryo death. American Journal of Botany 76: 6-13.

Cooper JL, Till BJ, Laport RG, Darlow MC, Kleffner JM, Jamai A, El-Mellouki T, Liu S, Ritchie R, Nielsen N, Bilyeu KD, Meksem K, Comai L, Henikoff S (2008) TILLING to detect induced mutations in soybean. BMC Plant Biology 8: 9.

de Ponti, OMB (1975) Breeding parthenocarpic pickling cucumbers (*Cucumis sativus* L.): Necessity, genetical possibilities, environmental influences and selection criteria. *Euphytica*25: 29-40.

Doerge RW (2002) Mapping and analysis of quantitative trait loci in experimental populations. *Nat. Rev. Genet.* 3: 43-52.

Elizabeta S, Lush S (2010) Induction of staminate flowers in gynoecious cucumber lines (*cucumis sativus* l.) Using silver nitrate. *Proceedings – 21st Scientific-Expert Conference of Agriculture and Food Industry*, Neum: 407-417

Ficcadenti N, Sestili S, Pandolfini T, Cirillo C, Rotino GL, Spena, A. (1999) Genetic engineering of parthenocarpic fruit development in tomato. *Mol Breed* 5: 463-470.

Fortescue JA, Turner DW (2005) The anatomy of ovule ontogeny of banana, plantain and ensete (*Musaceae*). *Scientia Hortic* 104: 479-492.

Garcia-Martinez JL, Hedden P (1997) Gibberellins and fruit development. In: FA Tomas-Barberan and RJ Robins ed. Phytochemistry of Fruit and Vegetables. Oxford: Clarendon Press. 577 p.

George W, Scott J, Spilttstoesser W (1984) Parthenocarpy in tomato. *Hortic Rev* 6: 65-84.

Gillaspy G, Bendavid H, Gruissem W (1993) Fruits - A developmental perspective. Plant Cell 5: 1439-1451.

Giovannoni JJ, Grandillo S, Martin GB, Messeguer R, Miller JC, Miller L, Paterson AH, Pineda O, Roder MS, Wing RA, Wu W, Young ND (1992) High-density molecular linkage maps of the tomato and potato genomes. *Genetics* 132: 1141-1160.

Gorguet B, Eggink PM, Ocana J, Tiwari A, Schipper D, Finker SR, Visser RF and vanHeusden AW. (2008) Mapping and characterization of novel parthenocarpy QTLs in tomato. *Theor Appl Genet* 116: 755-767.

Gorguet B, Van-Heusden AW, Lindhont P (2005) Parthenocarpic fruit development in tomato. *Plant Biol* 7: 131-139.

Gustafson FG (1942) Parthenocarpy: Natural and artificial. *Bot Rev* 8 (9): 599-654.

Hawthorn LR (1937) Seedlessness in tomatoes. *Science* 85: 199.

Huang JK, Pray C, Rozelle S. (2002) Enhancing the crops to feed the poor. *Nature* 418: 678-684.

Johnson SP, Hall WC (1954) Parthenocarpy in tomato. *Proc Amer Soc Hort Sci* 63: 329-332.

Kikuchi K, Honda I, Matsuo S, Fukuda M, Saito T. (2008) Stability of fruit set of newly selected parthenocarpic eggplant lines. *Sci Hort* 115: 111–116.

Kim IS, Yoo K, Okubo H (1994) Studies on parthenocarpy in *Cucumis sativus* L. V. Influence of exogenous plant growth regulators on growth and diffusible IAA level of cucumber ovaries. *J Kor Soc Hort Sci* **35**: 195-200.

Kojima K, Tamura Y, Nakano M, Han D-S Niimi Y (2003) Distribution of indole-acetic acid, gibberellin and cytokinins in apoplast and symplast of parthenocarpic tomato fruits. Plant Growth Regul 41: 99-104.

Koltunow AM, Grossniklaus U (2003) Apomixis: a developmental perspective. *Ann. Rev. Plant Biol.* 54: 547-574.

Kuno S, Yabe K (2005) Genetic analysis of parthenocarpy and spineless in the F_2 segregating. Res Bull Aichi Agric Res Cent. **37**: 29–33.

Lasley BL, Loskutoff NM, Anderson GB (19940 The limitation of conventional breeding programs and the need and promise of assisted reproduction in nondomestic species. *Theriogenology* 41: 119-132.

Ledbetter CA, Ramming DW (1989) Seedlessness in grapes. *Hortic Rev* 11: 159-184.

Lee TH, Sugiyama A, Takeno K, Ohno H, Yamaki S (1997) Changes in content of indole-3- acetic acid in activities of sucrose metabolising enzyme during fruit growth in eggplant (*Solanum melongena* L.). *J Plant Physiol* 150: 292-296.

Masuda M, Agong SG, Tanaka A, Shikazono N, Hase Y (2004) Mutation spectrum of tomato induced by seed radiation with carbon and helium ion beams. *Acta Hortic* 637: 257-262.

Meshcherov ET, Juldasheva LW. (1974) Parthenocarpy in cucumber. *Trudy-po-Prikladnoi- Botanike-Genetiki-I-Selektsii* **51**: 204-213.

More TA, Budgujar CD (2002) Isolation of parthenocarpic tropical gynoecious lines in cucumber (*Cucumis sativus* L.). *Acta Hortic* **588**: 255-260.

More, T. A. and Seshadri, V.S. (1988) Development of tropical gynoecious lines in cucumber. *Cucurbit Genet Coop Rep* 11: 17-18.

Pandolfini, T., Rotino, G. L., Camerini, S., Defez, R. And Spena, A. (2002) Optimization of transgene action at post-transcriptional level: high quality parthenocarpic fruits in industrial tomatoes. *BMC Biotechnol* 2: 1.

Pecaut P, Philouze J (1978) A *sha pat* line obtained by natural mutation. Tomato Genetics Cooperative Reports 28: 12.

Philouze J, Maisonneuve B (1978) Heredity of the natural ability to set parthenocarpic fruits in the Soviet variety Severianin. Tomato Genet Coop **28**: 12- 13.

Rotino GL, Perri E, Zottini M, Sommer H, Spena, A. (1997) Genetic engineering of parthenocarpic plants. *Nat Biotechnol* 15: 1398-1401.

Schwabe WW, Mills JJ (1981) Hormones and parthenocarpic fruit set: A literature survey. *Hort Abstr* 51: 661-698.

Soressi GP (1970) Tomato mutants following EMS seed treatments. Tomato Genetics Cooperative Reports: 20: 59.

Sugiyama K, Morishita M (2001) A new method for producing diploid seedless watermelon. *Acta Hortic* 588: 238-249.

Sugiyama K, Morishita M (2000) Production of seedless watermelon using soft-X irradiated pollen. *Scientia Hortic* 84: 255-264.

Sun Z, Lower RL, Staub JE (2006) Analysis of generation means and components of variance for parthenocarpy in cucumber (*Cucumis sativus* L.). Plant Breeding **125**: 277-280.

Tiwari A, Vivian-Smith A,Voorrips RE, Habets M, Xue L, RemkoOffringa R, Heuvelink E (2011) Parthenocarpic potential in *Capsicum annuum* L.is enhanced by carpelloid structures and controlled by a single recessive gene. *BMC Plant Biol* 11: 143-155

Tsao T (1980) Growth substances: Role in fertilization and sex expression. In: Skoog, F. (*Ed*.). Plant Growth Substances. Spring-Verlag, N.Y. pp.345-348.

Vanamala J, Cobb G, Loaiza J, Yoo K, Pike LM, Patil BS (2007) Ionizing radiation and marketing simulation on bioactive compounds and quality of grapefruit (*Citrus paradise* cv. Rio Red). *Food Chem* 105: 1404-1411.

Vivian-Smith A (2001) The molecular basis for the initiation of fruit development and parthenocarpy. PhD dissertation (Adelaide, Australia: University of Adelaide).

Yao JL, Dong YH, Morris BA (2001) Parthenocarpic apple fruit production conferred by transposon insertion mutations in a MADS-box transcription factor. Proceedings of the National Academy of Sciences of the United States of America 98: 1306-1311.

Yoshida T, Matsunaga S, Saito T (1998) Inheritance of parthenocarpic character in eggplant. J Japan Soc Horti Sci **67**: 257.

Ziglstra S. (1985) Parthenocarpie in tomaat; twee nieuwelijnenuitsoortkruising. Zaadbelangen 4: 92-94.

Chapter 8

Snap Melon (*Cucumis melo* var. *momordica* Duthie and Fuller): A Minor Cucurbit

Jyothi Motapalukula, Mallesh Sanganamoni, Vittal L. Mangi and Ankit Panchbhaiya

ABSTRACT

Snap melon (Cucumis melo var. momordica Duthie and Fuller) is one of the important cucurbitaceous vegetable crop. It has multiple uses as salad or cooked vegetable in case of immature fruits, matured fruits as desserts and seeds and seed oil also used for consumption. Besides these uses, snap melon exhibits huge diversity among the cultivars and has the cross compatibility with related species like musk melon can be exploited for improvement of valuable traits relating to quality, biotic and abiotic stress conditions. Cultivation of this crop provides the benefits among the farmers and also the consumers by providing adequate income and nutrition respectively. Though, this crop has importance but not so commercial among the farming community due to it's varied market acceptability.

Keywords: *Snap melon, Cultivation, Production technology, Momordica.*

Introduction

Scientific Classification

Kingdom	: Plantae
Order	: Cucurbitales
Family	: Cucurbitaceae
Genus	: *Cucumis melo*

Species : *C. melo var.* Momordica

Chromosome Number. : 2n=2x=24 (Singh and Roy, 1974)

The term "Melon" diversed in many different plants belongs to the family Cucurbitaceae (Genera: 1. *Momordica* 2. *Benincasa* 3. *Citrulus* 4. *Cucumis*). Snap Melon (*Cucumis melo var. momordica* Duthie and Fuller) is most favourite, important minor cucurbitaceous fruit vegetable grown throughout India for want of its seed kernel (Meena *et al.*, 2016), oil and fruit, where it is popularly called as "Phoot"(which split naturally at the time of ripening) (Maheshwari and Haldhar, 2018; Swamy, 2017; Dhillon *et al.*, 2014). Snap melon is an indigenous cucurbit from India (Swamy, 2017).

Common Names in different Languages

English	: Snap Melon
Sanskrit	: Alakakhya
Hindi	: Baro-kachro
Tamil	: Chukkarigai
Rajasthan/Marwari	: Kachra
Bengali	: Phuti
Telugu	: Karbooja dosa
Malayalam	: Kakkarikka-vitta
Marathi	: Chibunda
Kannada	: Kalangari

Distribution and Importance of the Crop

India is considered as the center of domestication of melon by some researchers with the earliest melon remains at the Indus Valley site of Harappa dated between

2300 and 1600 BC (Vishnu-Mittre, 1974). It is widely distributed from the place of origin to the different places of world. The major production belts are confined to Rajasthan, Punjab, Gujarat, Haryana, Uttar Pradesh, Bihar and West Bengal in India (Devendra *et al.*, 2013) and Southeast Asia, for instance in Myanmar and Vietnam are also cultivate this crop (Yi *et al.*, 2009). China is the world leading producer of snap melon (Devendra *et al.*, 2013). Thousand and above accessions of snap melon are maintained in four national gene banks in India namely National Bureau of Plant Genetic Resources, New Delhi; Central Institute for Arid Horticulture, Bikaner; Central Horticultural Experiment Station, Bhubaneswar; Punjab Agricultural University, Ludhiana, but they have not been comprehensively evaluated against various biotic and abiotic stresses (Dhillon *et al.*, 2014).

Snap melon is an underutilized, cross compatible species (Dhaliwal, 2017) of cucurbitaceous crop, exhibits tremendous diversity in morphological difference, fruit size, shape, colour, quality, taste and much diversity in germplasm (Tewari *et al.*, n.d.). Most of the snap melon accessions of Indian have been reported to be a good source of insect pest and disease resistance and also used as reference accessions throughout the world (Singh *et al.*, 2009; Cohen *et al.*, 2003). Snap melon has also been found to have resistance to downy mildew (More, 1999; Sambandam *et al.*, 1979).

Snap melon is a source of high fruit acidity, a trait that has been utilized to breed uniquely flavour melon cultivars (Dhillon *et al.*, 2014) with the help of landraces IC-274021 (0.61 per cent) and IC-267360 (0.57 per cent). The low-pH gene derived from Indian snap melon accession IND-35 was exploited using marker-assisted selection by Syngenta in 2008 to develop the pleasant-tasting F_1 hybrid melon 'GWANIPA' that was commercialized in UK, Germany and Holland by Kernel Export. This melon has a lemon flavour and contains 700-800 mg citric acid per 100 g fresh weight (F.W) with a pH level of 4.5 (Patent No. EP 1587933 B1) (Casanueva *et al.*, 2010). Fruit has an immense use for human being. Melons also provide edible oil and have various traditional medicinal applications.

Snap melon fruits are playing a very important role in nutrient supply (Table 5.1). It also contains good sources of antioxidants, carotenoids, niacin, vitamin A, B_6, C, potassium, iron, calcium and their high water content, make it an excellence diuretic (Meena *et al.*, 2016) and also help in promoting body hydration. Consuming moisture rich foods dissolves kidney stones (Singh *et al.*, 2004), keeping digestive system healthy, relieving constipation (Pareek *et al.*, 1999).

Table 8.1. Nutritional Composition of Snap Melon

Nutrient Content	*Per 100g Fruit Sample*
Moisture	79.93 g
Crude protein	0.369 g
Fat	1.12 mg
Carbohydrate	15.6 g
Crude fibre	1.34 g
Total ash	1.64 g
Calcium	0.76 mg

Nutrient Content	Per 100g Fruit Sample
Phosphorous	0.088 mg
Iron	0.843 mg
Zinc	0.2023 mg
Ascorbic acid	18.6 mg
Energy	74.0 kcal

Samadia, 2003.

Uses

Melons are usually consumed at fresh stage, raw, cooked, or used in juice, desserts, fruit salads, or custards. After removing the coat, seeds are used in bakery products and the traditional drink called as 'Thandai' (Goyal and Sharma, 2009). Fruits of snap melon are eaten either raw or cooked as a vegetable. Immature fruit is used as a pickling melon. Unripe dehydrated fruits locally called as 'Khelra' are also used in off season (Goyal and Sharma, 2009). The ripe snap melon fruits are used as dessert (Dubey *et al.,* 2016; Singh, 2013) or as salad also used to prepare rayata (Singh, 2013; Goyal and Sharma, 2009). The light pink flesh fruits are mealy, with an insipid taste/tasteless or slightly sour flesh. To enhance the taste of snap melon fruit pieces can also mix with other fruits. It can process into different value added products namely, Jam, pickles, chutney, syrup, squash and as dehydrated. Seed is rich in oil with a nutty flavour but very fiddly to use because the seed is small and covered with a fibrous coat. Edible oil is obtained from the seed. The seed is antitussive, digestive, febrifuge and vermifuge (Komor and Devi, 2016; Devendra *et al.,* 2013; Duke and Ayensu, 1985). The seed contains between 12.5 - 39.1 per cent oil (Muthuselvi *et al.,* 2019). Ripe fruits can be used to make drink mixed with jaggary or sugar. Seed kernels are extracted and extensively used in sweets, bakery products and drinks (Traditionally called as Thandai).

Fruits of snap melon can be used as a refreshing cooling effect, intestine cleanser or natural moisturizer for skin and first aid treatment for burns, abrasions (Devendra *et al.,* 2013). The fruit is stomachic (Meena *et al.,* 2016; Devendra *et al.,* 2013), appetite, relieving constipation. The flowers are expectorant and emetic (Meena *et al.,* 2016; Devendra *et al.,* 2013; Duke and Ayensu, 1985). Seeds were considered to possess antioxidant, anti-inflammatory and analgesic value (Milind and Singh, 2011). The sprouting seed produces a toxic substance in it's embryo (Rana and Brar, 2017).

Botany

Cucumis melo L. *var. momordica* is an annual climber growing up to 1.5 m (5ft) (Brar and Rana, 2018), able to climb with the help of tendrils, but more commonly sprawling along the ground (Komor and Devi, 2016). It shows sufficient variability in bearing habit, sweetness and fruit morphology (Rakesh *et al.,* 2016). The vine is monoecious and self-fertile (Swamy, 2017; Dhillon *et al.,* 2014). Individual flowers are either male or female, but both sexes can be found on the same plant and pollination take place by insects. After which only female flowers are advances

Vine

Male Flower

Female Flower

Ripened Fruit

into fruits. Abundance production of male flowers can be seen during early stage of flowering. However, time being both male as well as female flowers are produced almost equal proportion (Tewari *et al.*, n.d.). The fruits vary in size from small to large, usually smooth and divergent shapes *viz.*, oblong, round, oblate, acorn, long, ovate, oblong spherical, spherical, elongated, cylindrical, pyriform and elliptical. Fruit flesh color varies from cream and yellow to orange 4 (Dhillon *et al.*, 2014) with a mealy, somewhat insipid or slightly sour flesh (Komor and Devi, 2016). The seed is small and covered with a fibrous coat.

Crop Production Aspects of Snap Melon

Improved Varieties

Pusa Shandar (Selection)

First commercial variety of snap melon developed at IARI, New Delhi. Oblong fruits with 700 g weight, medium size, creamy white skin with thick and light pink flesh. Early maturing variety (45-50 days). Tolerant to many disease. Average yield 380 q/ha, an increase of 32 per cent over best local check and used for cultivation purpose as well as home or kitchen garden.

Konkan Madhur (Mass Selection)

It is high yielder (15-31 t/ha) and less cracking (9 per cent) variety. The fruits of this variety have firm light orange colour pulp and good keeping quality (3-4 days). It is tolerant to pest and diseases. It is recommended for cultivation in Konkan region during Kharif season.

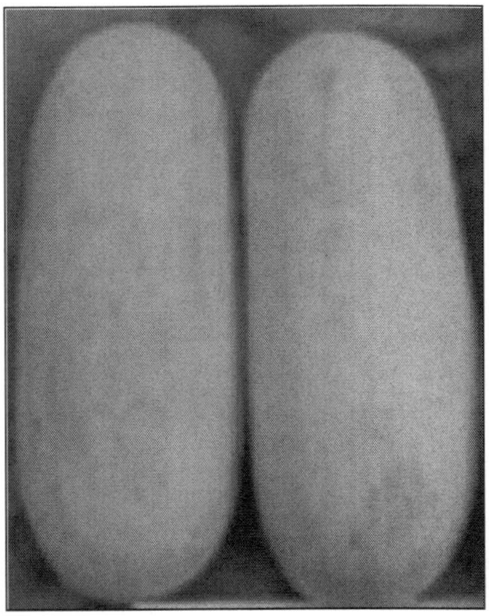

Pusa Shandar

Konkan Madhur

AHS-10 (Selection)

Developed at CIAH, Bikaner (Rajasthan). Oblong fruits with 900 g weight, 17-20 cm long and 9.7-10.5 cm diameter. Maturity is in 68 days. Average yield is about 200-220 q/ha. Whitish pink flesh and sweet in taste (TSS 4.5-5 °Brix)

AHS-82 (Selection)

Developed at CIAH, Bikaner (Rajasthan). Fruits are oblong with 900 -950 g weight, 22-24.5 cm long. Maturity is in 67-70 days. Average yield 225-248 q/ha. Flesh is whitish pink and sweet in taste (TSS 4.3- 4.9 °Brix).

Shantiniketan SM-1 (Selection)

This variety is suitable for the red and lateritic belt of West Bengal, India. Fruits are less sweet in taste, TSS 4.52°Brix. Average yield is about 106.52 q/ha.

Climatic Requirements

Snap melon is a warm season vegetable crop mainly grown in tropical, sub-tropical regions of the world and also grown to limited portions of the temperate regions, where there is relatively long summer. This crop is relatively drought tolerant to a large extent and survive well in the regions with low rainfall and hot climate conditions (Pareek and Samadia, 2002; Tewari *et al.*, n.d.). It can be grown at elevations up to 1000 m mean sea level. They prefer a mean annual rainfall in the range of 1000 mm to 1300 mm, but it can tolerate 900 and 2500 mm both lower and higher sides, respectively. The conditions reduce fertilization and lead to diseases like too much rainfall or too high a humidity. The optimum annual day time temperature range is between 18°C and 30°C, though it tolerate 9°C and 35°C at lower and higher limits respectively. Seed germination can be successful between the temperature range of 18°C to 25°C. It is sensitive to frost as it is classified under tender annual plant and even cannot grow in the shade. As a basic rule in case of all cucurbits the short day, low temperature increases female flowers and very high temperature, long day conditions increases male flowers as well as flower drop.

Soil Requirement

Snap melon suitable for growing under wide range of soils ranging from light sandy, medium loamy and heavy clay soils. The only requirement for better productivity is organic rich, well-drained moisture retentive soil. Crop is relatively sensitive to acidic soils and preferable pH range is 6.5 to 7.5 (Rana and Brar, 2017) but it can tolerate as less as pH 5.0 and as more as pH 8.7.

Propagation

Snap melon is propagated through seed having viability up to one year in ambient condition. Seed can be sown in-situ or in containers or protrays. Seed will germinate within two weeks. When using containers in the areas prone to high frost conditions, sow two or three seeds per pot and thin out to the best seedling. Grown seedlings of about 15cm tall are transplanted in main field after ending of peak frost.

Field Preparation

Snap melon is usually raised on ground though it is a vine crop because of it's heavy fruit weight. To have more yields with better quality fruits raised beds are preferred for sowing. In areas where the water availability is a problem, ridge and furrow method is followed for growing. In case of river bed cultivation shallow pits are prepared with 30 cm x 30 cm x 30 cm dimension of length, width and depth, respectively at the spacing of 1.5 to 2.0 meters.

Sowing Season

Sowing season of snap melon varies from region to region. Mostly it is done during early to mid spring. In Goa, Konkan areas of Maharashtra and Karnataka sowing is done in June to July. While Rajasthan and West Bengal sowing take up in January to March. In case of Kerala and Tamil Nadu it is Feb- May.

Seed Rate and Spacing

Seed of snap melon is comparatively small as compared with other cucurbits. Seed rate for sowing of the crop varies from 2-3 kg per ha. Seeds are sown at the spacing of 75 to 100 cm between plant to plant and row distance may extend up to two meters.

Irrigation Management

First irrigation is applied immediately after sowing. The frequency of irrigation varies from the season to season. In summer season four to five irrigations are required and during rainy season it may or may not required depending on the rains. Vine growth, flowering and fruiting stages are critical stages of crop growth for irrigations.

Nutrition Management

Nutrient requirement varies with the soil pH, soil type, growing conditions, irrigation facilities, variety or hybrid to be grown. Well decomposed farm yard manure of 20-25 ton per hectare is applied at the time of field preparation. Need based fertilizer application can be done to avoid excess application of nitrogen, as it leads to luxurious plant growth. In general snap melon is not a heavy feeder of nutrients. About 40 kg nitrogen is applied in two equal splits including basal application. Phosphorus and potassium is applied based on the need of the soil.

Weed Management

Snap melon demands good weed control during early stages of crop growth up to 30 to 45 days after sowing. For this purpose two shallow hoeings are required. In later stage crop will control the weeds by spreading and covering the ground area.

Pest and Disease Management

Insect pests like fruit flies, red pumpkin beetle and leaf minor are usually causing the damage in spine gourd. Powdery mildew, downy mildew, angular leaf spot, alternaria leaf blight and anthracnose are common diseases in snap melon.

Pest

Fruit Fly (*Bactrocera cucurbitae*)

Adult flies lays eggs beneath the surface of the fruits, when it is under development. As soon as eggs hatches maggots causes damage by feeding on the pulp of the fruits and can be identified by oozing of resinous fluid from fruits. Due to this, fruits will become distorted and malformed finally leads to premature dropping. Such fruits are unfit for marketing and consumption.

Management

Collective operations are required to manage this pest *viz.* collect infested and fallen fruits and burn in deep pits, change the sowing date as the fly population is low in hot dry conditions and at it's peak during rainy season, expose the pupae by ploughing and turning over soil after harvest and use ribbed gourd as trap crop and apply Emamectin Benzoate (5 per cent SG) at 0.4-0.5 g or Acetamiprid (20 per cent SP) at 0.1 g or Carbosulfan (25 per cent EC) at 1-1.5 ml per liter of water on congregating adult flies on the undersurface of leaves. Use attractants like citronella oil, eucalyptus oil, vinegar (acetic acid), and lactic acid to trap flies. Use of baits *viz.* Peel of oranges or cucumbers or their fruit pulp, 100 ml ammonia or cow urine, 0.5 litre of water; all mixed well and allowed to stand overnight. For ready-to-use bait, mixture is diluted with 15 litre of water. In another method 6 ml yeast extract, 0.5 g sodium sulphite (Na_2S), one litre of water is used (Stoll, 2000). Insecticides can also be used in a mix with attractants like cue lure and methyl eugenol. This is a technique called Male Annihilation Technique (MAT) and consists of many bait traps (16 traps) throughout the field. The mixture can be applied in traps or other devices. Insecticides can also be mixed with protein bait, which controls both, male and female fruit flies. A hydrolysed protein insecticide mix is applied on spots (spot technique) on the crop canopy (Anon., n.d_a). Spraying the soil under trees or crops with Metharhizium anisopliae or other entomopathogenic fungi may be helpful in reducing fruit fly larvae in the soil. It also requires that the fungal product used exhibits high effectiveness against fruit fly (Anon., n.d_b).

The accessions IC- 430190 (11.21 per cent), DKS-AHS 2011/4 (14.97 per cent) and DKS-AHS 2011/3 (18.57 per cent) were found to be novel resistant accessions against melon fly, *B. cucurbitae* infestation (Haldhar *et al.*, 2018).

Red Pumpkin Beetle (*Aulacophora foviecollis*)

This beetle feeds on the foliage of the vine and makes holes on the leaf. In severe infestation total plant will be defoliated and effects photosynthesis there by productivity of the crop.

Management

1. Collection and destruction of adult beetles
2. Removal of alternate host from field and bunds
3. The predator reduviid bug, *Rhinocoris fuscipes* (Fab.) is a larval parasites can be released for parasitization (Rajagopal and Trivedi, 1989).
4. Spraying with effective insecticides like Chlorpyriphos (20 per cent EC) at 2 ml or Profenofos (50 per cent EC) at 2 ml or Emamectin Benzoate (5 per cent SG) at 0.4 g per liter of water.

Leaf Minor (*Liriomyza* spp.)

The adult fly lays eggs inside the leaves and larvae feed inside the leaf making tunnels with serpentine mines. In later stage drying and dropping of leaves due to severe infestation can be seen.

Management

1. Collection and destruction of infested leaves
2. Removal of alternate host from field and bunds
3. Spray Neem Seed Kernel Extract (NSKE) at 3 per cent concentration
4. Mildews (powdery mildew and downy mildew)

The first sign of powdery mildew is pale yellow leaf spots. White powdery spots can form on both upper and lower leaf surfaces, and quickly expand into large blotches. The large blotches can cover entire leaf, petiole and stem surfaces. When powdery mildew infects the majority of the foliage, the plant weakens and

Powdery Mildew
(*Erysiphe cichoracearum*)

Downey Mildew
(*Pseudopernospora cubensis*)

the fruit ripens prematurely. While, in downy mildew lesions are formed and that are limited by the leaf veins. During periods of leaf wetness from dew, irrigation or rainfall, incipient lesions can become conspicuously water-soaked. This is the earliest symptom produced by the disease, but will disappear as moisture dissipates. Early lesions are light green in appearance and become chlorotic and finally necrotic as host plant cells die. Severe infection results in leaves that are completely dead and curled up. This symptom has been described as "wildfire" as the leaves appear to be burned.

Management

1. Avoid excess nitrogen application
2. Spraying with fungicides *viz.* Azoxystrobin (23 per cent SC) at 0.6 g or Chlorothalonil (75 per cent WP) at 1.5 g or Hexaconazole (2 per cent SC) at 2.5 ml or Propiconazole (25 per cent EC) at one ml per liter of water will serve the purpose.

Angular Leaf Spot (*Pseudomonas lachrymans*)

This pathogen causes chlorosis and necrosis of leaf tissue. In later stages lesions progress and may turn yellow and then brown and finally the leaf tissue may tear.

Management

1. Spraying with streptocycline (0.05 per cent) at 0.50 g per liter of water along with copper oxy chloride (COC, 50 per cent WP) at 2.5 g per liter of water will serve the purpose effectively.

Alternaria Leaf Blight (*Alternaria cucumerina*)

Characteristic symptoms first appear on leaves as small, circular and light brown to reddish brown spots of different size which, later enlarge in a concentric pattern or rings and margins appear. These spots coalesce to form larger necrotic areas on leaves.

Management

1. The infected crop debris and weeds collected and burnt
2. Use of disease free and certified seed.
3. Sprayings of mancozeb (75 per cent WP) at 2g per liter or Azoxystrobin (23 per cent SC) at one ml per liter in regular interval was very effective against this disease.

Anthracnose (*Colletotrichum lagenarium* and *Colletotrichum orbiculare*)

The symptoms of this disease are appear as small yellowish or water soaked areas on leaves which, enlarge rapidly and turn brown. If the infection on fruit pedicle, the young fruit may be darken, shrivel and dry. Circular black spots or cankers appear on fruits. On maturing fruit, lesions appear as small, circular, sunken areas which may grow to the size of a quarter or larger on melons based on the host and environmental conditions. In the sunken black spots, salmon pink colored masses of spores can be seen under humid conditions.

Management

1. Proper drainage in the field.
2. Proper distance between the plants should be maintained so as to reduce dense canopy.
3. Practice crop rotations with unrelated crops for at least two years.
4. Follow good sanitation practices, such as cleaning up crop debris at the end of the growing season.
5. Spray the crop with copper oxy chloride (COC, 50 per cent WP) at 2.5 g per liter of water carbendazim (50 per cent WP) at 2 g per liter and repeat at 7 to 10 day intervals if necessary.
6. Treat the seeds with carbendazim (50 per cent WP) at 2 g per kg seed
7. Spray the crop with chlorothalonil (75 per cent WP) at 1.5 g per liter of water

Use of Plant Growth Regulators

Male and female flower production can be altered by various factors *viz.*, temperature, photoperiod, nutrition and by the application of plant growth regulators. Exogenous application of plant growth regulators at two to four true leaf stage, can alter the sex ratio and sequence. Two to four true leaf stage is the critical stage at which the suppression or promotion of either sex is possible. Treatment with auxin increases the female sex tendency. While gibberellins cause a shift towards maleness at higher concentration and femaleness at lower concentration.

Harvesting and Yield

Fruits are ready to harvest within 60-70 days after sowing. Harvesting is mainly done when fruits attain full size and still in tender for pickle and culinary purpose. Harvesting is done at the interval of three to four days in the peak season. In case of table purpose harvest fruits when abscission layer is fully developed. Fruit is easily separated from the peduncle and colour changes pale yellow to orange. At over maturity time fruit always cracks/split/explode and scattering their seeds. Fruit cracking is either longitudinal or starting in the middle of fruit, though in some instances only skin peeling (longitudinal or random) occurs. Hence, it is called as a *'phoot'* or *'snap melon'* or *'phoot kakar'* or *'kakadia'* or *'phut'* which means to split.

An individual plant is usually allowed to bear four melons. The yield varies from 10 to 20 ton per ha, with an average of 11 - 13 ton per ha and seed yield varies from 1 to 3 q per ha.

Postharvest Management and Marketing

Snap melon fruits are highly perishable, they have poor storage life. These cannot keep well so have to send market immediately after harvesting. Fruits can be store for 2-4 days at room temperature. In cold storage conditions fruits can be store two to three weeks at 2°C to 4°C and 85-90 per cent RH. Due to highly perishable nature of snap melon fruits, to prolong it's utilization, those fruits can be processed into different value added products and can consume during off-season *viz.*, candies, ice creams, Dehydrated slice/pieces (Khelra), pickles, jam, chutney, squash, Syrup. Snap melon fruit syrup can be prepared by adding sugar with pulp in the ratio of 1:2, citric acid (at different acidity levels) and potassium metabisulphate at 700 ppm as preservative. Such syrup can be stored up to three months at ambient temperature.

REFERENCES

Anonymous n.d$_a$ German Agency for International Cooperation (GIZ). (n.d.). Fruit fly Management. pp6.V. https: //www.asean-agrifood.org/?wpfb_dl=160.

Anonymous n.d$_b$ German Agency for International Cooperation (GIZ). (n.d.). Fruit fly Management. pp7. https: //www.asean-agrifood.org/?wpfb_dl=160.

Brar NS, Rana MK (2018) Snap melon, Vegetable crops science, CRC Press Taylor and Francis group New York, pp.476.

Casanueva A, Foncelle B, Nicolet JE, Van Doon JE, Oliver SM (2010) Novel melon plants. Patent no. 20100034952. http: //www.faqs.org/patents/app/20100034952.

Cohen Y, Meron I, Mor N, Zurial, S, (2003). A new pathotype of *Pseudoperenospora cubensis* causing downey mildew in cucurbits in Israel. *Phytoparasitica* 31: 452-466.

Devendra K, Ram RB, Sanjay K, Sutanu M, Manoj K, (2013) Variability and physico-chemical studies in snap melon (*Cucumis melo var. momordica*). *Asian J of Horti* 8(**2**): 751-753.

Dhaliwal MS, (2017) Cucurbits. Handbook of Vegetable Crops, 3rd edn. Kalyani Publishers, pp.77- 147.

Dhillon NP, Singh H, Pitrat M, Monforte AJ, Mc Creight JD, (2014) Snapmelon (*Cucumis melo* L. *Momordica* group), an indigenous cucurbit from India with immense value for melon breeding. In XXIX *Int. Horticultural Congress on Horti*: Sustaining Lives, Livelihoods and Landscapes (IHC2014): 1102 : 99-108.

Dubey RK, Muddarsu VR, Singh V, Upadhyay G, Singh S, Pandey AK, Pandey s, Rao GP, (2016) Studies on seed protein profiling in snap melon (*Cucumis melo var. momordica* L.) Genotypes of Northeast India. *Int J Agricult Stat Sci* 12(**1**): 143-149.

Duke JA, Ayensu ES, (1985) Medicinal Plants of China, Reference Publications, Inc.

Goyal M, Sharma SK, (2009) Traditional wisdom and value addition prospects of arid foods of desert region of North West India. *Indian J Traditional Knowledge* 8(**4**): 581-585.

Haldhar SM, Samadia DK, Bhargava R, Choudhary BR, Singh D, (2018) Host plant accessions determine bottom-up effect of snap melon (*Cucumis melo var. momordica*) against melon fly (*Bactrocera cucurbitae* (Coquillett)). Breeding Science, 68(5): 499-507.

Komor PD, Devi OS, (2016) A edible bio-resources and livelihoods. Assam State Biodiversity Board Guwahati. pp62-63.

Maheshwari SK, Haldhar SM, (2018) Disease management in arid horticultural crops. CIAH/Tech./Pub. No. 68, ICAR-Central Institute for Arid Horticulture, Bikaner, Rajasthan, India, pp 35.

Mc Cormack J (2005) Cucurbit seed production: An organic seed production manual for seed growers in the Mid-Atlantic and Southern US Creative Commons, 559 Nathan Abbott Way, pp20.

Meena SR, Singh RS, Sharma BD, Singh D, (2016) Most favourite traditional cucurbitaceous vegetables and their utilization pattern in Thar desert of the western Rajasthan, India. *Indian J. traditional Knowledge* 15(**3**): 385-394.

Milind P, Singh K (2011) Muskmelon is eat-must melon. *Int J Pharmacy*, IRJP 2(**8**): 52-57.

More TA, (1999) Breeding muskmelon (*Cucumis melo*. L) resistance to diseases. In: Souvenir and Abstr. National Symposium on crop pest management. Challenge for Next Millenium. Nov. 16. MPKV, Rahuri, India, pp27-28.

Muthuselvi R, Praneetha S, Rajasree V, Kumaravadivel N, (2019) Evaluation of snap melon (*Cucumis melo var. momordica* Duth. and Full) genotypes for growth, yield and quality parameters, *Int J Chem Studies* 7(**5**): 364-367.

Pareek OP, Vashistha BB, Samadia DK, (1999) Genetic diversity in drought hardy cucurbits from hot arid zone of India. IPGRI Newslett. Asia Pacific Oceania, 28: 22-23.

Pareek OP, Samadia OK, (2002) Promising indigenous cucurbit varieties. *Indian Hort* July-Sept. pp.15–18.

Rajagopal D, Trivedi TP, (1989) Status, bioecology and management of Epilachna beetle, *Epilachna vigintiocto punctata* (Fab.) (Coleoptera: Coccinellidae) on potato in India: A review, Tropical Pest Management, 35: 4, 410-413, DOI: 10.1080/09670878909371418.

Rakesh K, Dubey VR, Muddarsu, Vikas S, Upadhyay G, Singh S, Pandey AK, Pandey S, Gangadhara RP (2016) Studies on seed protein profiling in snap melon (*Cucumis melo var momordica* L.) Genotypes of northeast India. *Int J Agricult Stat Sci* 12(**1**): 143-149.

Rana MK, Brar NS (2017) Snap melon, Rana, MK (ed.). Vegetable crop science. CRC Press

Samadia DK (2003) Snap melon production technology for arid and semi arid areas. *Intensive Agri*, 40(**1-2**): 27-29.

Sambandam CN (1979) Inheritance of Certain Qualitative Characters in Musk Melon (Cucumis *melo* L.) *Annamalai Univ Agric Res* 9: 41-44.

Singh AK, Roy RP (1974) Karyological studies in *Cucumis* (L.). Caryologia, 27(**2**): 153-160.

Singh HP, Rai M, Pandey S, Kumar S (2009) Vegetable Varieties of India. Studium Press (India) Pvt. Ltd., New Delhi, p. 325.

Singh JP, Beniwal RK, Yadava ND, (2004) Ethno botanical importance of cucurbits in India. Advances in Arid Horticulture: Present Status, edited by Saroj PL, Vashishtha BB and Dhandhar DG, 1: 203-219.

Singh SP (2013) Cucurbits biodiversity, breeding, and production in Uttar Pradesh. 1st edt. Uttar Pradesh State Biodiversity Board, Lucknow, India pp2.

Stoll G (2000) Natural Crop Protection in the Tropics. Margraf Verlag, Weikersheim, Germany.

Swamy KRM (2017) Origin, distribution and systematics of culinary cucumber (*Cucumis melo subsp. agrestis var. conomon*). *J Hort Sci* 12(**1**): 1-22.

Tewari P, Seena M, Singh YV, and Roy MM, Kachra (*Cucumis melo*): An under Utilised Gift of Nature, Director, CAZRI, Jodhpur. http: //www.cazri.res.in/ publications/KrishiKosh/167-(KACHRA).pdf

Vishnu-Mittre (1974) Palaeobotanical evidence in India. In Evolutionary Studies in World Crops, Diversity and Changes in Indian Subcontinent, *J. Hutchinson, ed. Cambridge*, UK Cambridge Univ. Press pp3–30.

Chapter 9

Spine Gourd (*Momordica dioica* Roxb. ex Willd.): An Indigenous Cucurbitacous Vegetable Crop

Mallesh Sanganamoni, Vittal L. Mangi, Ankit Panchbhaiya and Jyothi Motapalukula

ABSTRACT

Introduction

Scientific Classification

Kingdom	: Plantae
Order	: Cucurbitales

Family : Cucurbitaceae

Genus : *Momordica*

Species : *M. dioica*

Spine gourd (*Momordica dioica* Roxb. ex Willd.) is a minor cucurbitaceous and dioecious perennial vegetable crop native to Asia with chromosome number 2n=2x=28. Although this genus *Momordica* is originated from Indo-Malayan region (Hooker, 1961). It is mostly grown as wild form but there is gradual expansion of cultivated area due to its enormous potential as vegetable by supplying plenty of nutrients as well as medicinal compounds for maintaining healthy diet. Spine gourd is grown mainly for want of its immature fruits, even young twigs and leaves are also used as vegetable (Bandyopadhyay and Mukherjee, 2009). Though, the spine gourd crop is already under cultivation, it is confined to lesser extent, as there are multiple drawbacks for cultivation of this crop at farmer level. Those drawbacks start from varietal availability up to the marketing. These limitations are responsible for not commercializing the spine gourd even it has huge consumer acceptance in the market. Because of this low availability in the market, price of the spine gourd fruits always fetches more throughout the season.

Vernacular Names of Spine Gourd

English : Spine gourd

Sanskrit : Karkotaki

Hindi : Kikoda

Tamil : Mezhuku-pakal, pazhu-pakal

Oriya : Kankada

Assamese : Bhat-kerela

Manipuri : Karot

Bengali : Kakrol, Ghi korola

Telugu : Boda kakara

East coast of Andhra : Aa-kakara-kaya

Gujarati : Kantola

Marathi : Kantole

Mizo : Maitamtawk

Kannada : Mada haagala kaayi

Chhattisgarhi : Kheksi

Distribution and Importance of the Crop

Spine gourd is majorly confined to Asian countries *viz.* India, Bangladesh, Nepal, Bhutan and Srilanka. In India, it grows widely in costal parts starting from Tamil Nadu, Andhra Pradesh, Odisha up to north eastern states namely Assam,

West Bengal, Meghalaya, Tripura, Manipur, Mizoram and Nagaland. This crop is also grows as wild in certain parts of central India *viz.* Telangana, Maharashtra, Chhattisgarh, Karnataka and other parts of the country.

M. dioica is a nutritionally superior vegetable crop contains 84.1 per cent moisture along with 7.7 g carbohydrate, 3.1 g protein, 1 g fat, 3.0 g fiber, 33 mg calcium, 42 mg phosphorus, 4-6 mg iron, 162 mg carotene, 0.05 mg thiamin, 0.16 mg riboflavin, 0.06 mg niacin and 52 calories of energy per 100 g edible portion of the fruit (Aberoumand, 2011). It also contains fair amount of ascorbic acid. Spine gourd plant also possesses some important medicinal properties which are sex specific (Sastri, 1962). The leaves of female *M. dioica* are used as an aphrodisiac, to eradicate intestinal parasites, and to cure fever, constipation, asthma, hiccups, and piles (Kumar and Prajapati, 2003). The root of female *M. dioica* is used in the form of a medicinal paste to heal bleeding piles, also used for benefit in headaches, kidney stones, and jaundice (Jadeja *et al.*, 2006, Talukdar and Hossain, 2014 and Oudhia, n.d). The fruit is considered as pungent, non-bitter, hot, alexiteric, stomachic, and laxative, and plays a role in cures for biliousness, asthma, leprosy, bronchitis, fever, tumors, urinary discharges, excessive salivation, and heart disease. Juice of the fruit is a domestic remedy for inflammation. Powder obtained from dried fruit is used to induce sneezing, leading to nasal clearing (Bairava and Rana, 2017). Ethanol extracts of fruits are used to protect and heal the kidneys. Fruits are also used in ulcers, piles, sores and obstection of liver and spleen. The unripe fruits work as appetizer astringent in barrels (Ram, 2004). The seeds are used for chest problems and stimulate urinary discharge (Khulakpam, 2015). The root of male *M. dioica* is used in the form of a paste to heal ulcers caused by snake bites (Satyavati, 1987 and Anjaria *et al.*, 2002). It is also useful to cure elephantiasis (Kirtikar and Basu, 1999). The male plant is pungent, hot, with wound-healing properties, and it targets diseases of the blood, eyes, and heart (Talukdar and Hossain, 2014). Apart from this, *M. dioica* is effective against acute renal failure (Jain and Singhai, 2009). Ethanol and aqueous extracts of *M. dioica* shows anti-oxidant and hepato-protective activity (Talukdar and Hossain, 2014). The ripe fruits are sweet, oily and laxative. The roots contain a tri-terpenoids saponin, which give water and can be used as a substitute for soap (Rajasree *et al.*, 2009)

Overall spine gourd performs some curative roles as anti-diabetic, anti-cancer, anti-fertility, abortifacient, anti-inflammatory, anti-oxidant, snake bites (Ram, 2004 and Khulakpam, 2015), scorpion sting, jaundice and bleeding pile properties.

Uses of Spine Gourd

Various plant parts are consumed in a variety of ways, *viz.* the delicious young tender fruits used for culinary purpose, as sole vegetable and sometimes mixed with other vegetables, meat, fish and prepares. Often, it can be fried with oils and used. Occasionally fruits are cooked with hot spices and prepared as good vegetable curry. Young leaves, flowers and seeds are also edible (Ram *et al.*, 2001 and Rai *et al.*, 2012). Spine gourd became more famous due to its bitter less taste with high nutritive, medicinal values and can easily blended with any other vegetables and meat.

Botany and Floral Biology

Female Flower

Male Flower

Tubers

Fruits

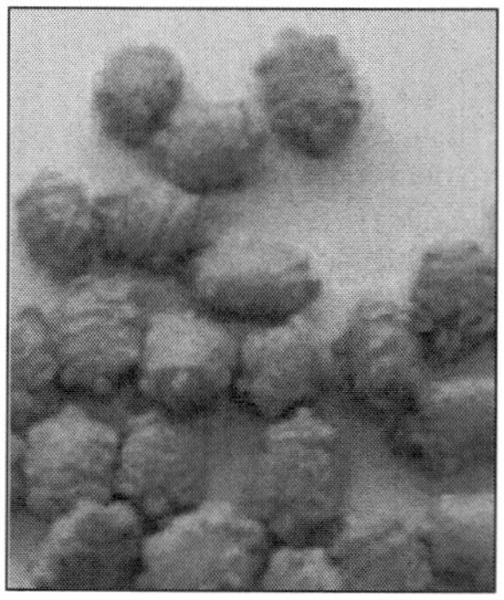

Seeds

Domesticated spine gourd plants are perennial crop with tuberous roots having dioecious in nature. Mostly single mother tuber is produced without any tabulates with varying sizes up to 500 to 600 g in weight. Tubers undergo into dormancy during winter. These plants are climbers with branched stems. Plant used to climb with help of tendrils produced at the axils of leaf in both male and female vines. Vines used to bear small leaves broadly ovate and membranous and three to five lobed (Baratakke *et al.*, 2013). In every axil of both male and female plants produces a single flower bud. In case of male flower bud takes 18-20 days from initiation to full blooming, while the female flower bud takes about 15 days. The male flower bud pedicle is usually elongated and takes the cup shape (Sandilya *et al.*, 2019). As soon as perianth leaves separated from each other the bud will open and releases the pollen grains. Both male and female flowers are appears yellow in colour after full opening (Sandilya *et al.*, 2019; Duthie, 1965; Khare, 2004; Kirtikar and Basu, 1999; Nadkarni, 2004; Oommachan, 1977 and Chopra *et al.*, 1956). Anthesis is taken place during late evening at 8:30 P.M. at the temperature range of 26.8°C to 27°C. The dehiscence started with the appearance of a slit on the anther lobe which resulted into liberation of creamy mass of pollen grains on the surface of anther. The dehiscence starts four and a half hour before anthesis and individual flower takes about 45 minutes to complete the process. Maximum release of pollen grains happens at 2:45 to 4:15 P.M. The receptivity of stigma lasts up to 24 hours. The stigma receptivity remained at a peak for 6-9 hours after opening of female flower (Dubey, 2006). Fruits are usually appears round to oval in shape with dense soft spines present on the surface of the fruits. The color of immature fruits is varying from light green to dark green and upon maturity these turn to yellow or orange. The ripened fruits split in three pieces and exposes the seeds with smooth bright red coated flesh (aril) on the surface. Individual fruit contains 10-20 seeds. Spine gourd seeds are usually contains hard seed coat, which creates the problem in germination.

Crop Production Aspects of Spine Gourd

Improved Varieties

Arka Neelachal Shree (Selection)

High yield, good appearance, high market preference.

Arka Neelachal Shree

Arka Neelachal Shanti (Hybrid)

It is developed through hybridization between spine gourd and teasel gourd, naturally pollinated, high yielding (15-16 kg/vine) with medium sized fruit (20g), moderately tolerant to fruit borer, anthracnose and downy mildew. It was released by SVRC, Govt. of Odisha in 2016. The variety exhibits plant and flower morphology more similar to teasel gourd while its fruit morphology is more close to spine gourd.

Indira Kankoda I (RMF 37) (Hybrid)

It is resistant to major pest and diseases. The plants start yielding after 75-80 days from seeds in the first year and 35-40 days from tubers from second year onwards up to six years. The green fruits are very attractive, dark green with long shelf life and single fruit weight is about 14 gram. The average green fruit yield is about 8-10 q/ha in the first year, 10-15 q/ha in the second year and 15-20 q/ha in the third year.

Arka Neelachal Shanti

Indira Kankoda I

Besides these improved varieties there are some local strains which are also available for cultivation among the farmers *viz.* Local Kantola, Ambika13-5, Ambika13-6, NDM-2, AJSG-1 and RMDSG-3.

Climatic Requirements

Spine gourd requires warm humid conditions for its maximum productivity. It can be grown both in tropical and sub-tropical parts with optimum temperature of 27°C to 33°C for better growth and development (Ponnusamy and Arumugam, 2019). More sun shine hours are required for better productivity of this crop. The well distributed rainfall of 900 to 1500 mm is ideal for rainfed production (Bharga, 2017). Existence of spine gourd can be seen up to an altitude 1500 meters mean sea level (Jadeja *et al.*, 2006). Spine gourd is frost sensitive and cannot withstand high winds. Crop will go into dormancy during winter season.

Soil Requirements

Spine gourd is a perennial crop with underground tuberous roots demands loose soils with sufficient organic matter. Compact soils are not suitable for its production. The ideal pH of 5.5 to 7.0 with loamy soil gives better performance of the crop (Anjana, 2019). Soil must be well drained and free from water stagnation, as the tubers are much susceptible to water logging conditions.

Field Preparation

Preparation and layout of the field is important aspect in spine gourd production. Selected land should be cleaned and marked with spacing of 1.5 m to 2 m between two rows and 0.5 m to 1 m between the plants. At the time of layout of the field clear marking should be done randomly for accommodating 10-15 per cent of compatible male plants to enable pollination, as spine gourd is a dioecious crop. It is also important to establish a supporting structure at the time field preparation as spine gourd is raised with the help of support formulated with any of the method to avoid touching of the vines to the ground. This supports are useful for getting quality fruits along with high yields. Most commonly followed method for supporting the crop is by constructing individual structure with GI iron pole of 1.5 m to 2 m height on which a frame work is made for spreading of the vines.

Propagation

Spine gourd can be multiplied with both sexual (seed) as well as asexual (vegetative) propagation methods.

Seed (Sexual) Propagation

Seed propagation is gaining importance as compare with other methods as it is easy to raise the seedlings by treating with any chemical substances for enhancing the germination percentage. For this purpose a seed rate of 1.5 kg to 3 kg is used to sow one hectare land area. In general two seeds are placed per hill which enables the thinning of male plants by retaining 10-15 per cent after reaches to the flowering stage. As this un-expected segregation of male and female plants are not possible to identify at seed or seedling stage the plants. This is the major drawback with seed propagation of spine gourd. It makes increased seed rate and adds additional cost of production and also wastage of resources. This problem can be rectified with the help of molecular marker technology at early stage of the crop development but it is very difficult to operate at farmer level. Even there is a problem with germination of the seeds due to its hard seed coat, which makes poor germination percentage of the seeds. It can be overcome by soaking the seeds overnight in the water or by treating the seed with chemicals like sulfuric acid. Lesser seed viability is also an issue in spine gourd seed propagation.

Vegetative (Asexual) Propagation

1. Tuber and Tuber Cuttings

Whole tuber or part of the tuber is used for planting to raise the crop. Though, it is convenient method for planting, not commercial due to more number of tubers requirement. Problem in tuber unavailability is because; individual plant produces only one mother tuber but no more tuberlets in this crop. It increases the cost of cultivation of spine gourd.

2. Vine Cuttings/Stem Cuttings

Propagation through vine cuttings is not so common method due to low rate of success and also requires hormonal treatment. In general 15–20 cm vine cuttings with 5-6 nodes from matured vine are used for multiplication.

3. *In vitro* Propagation/Micro-propagation

This micro-propagation procedure could be useful for raising genetically uniform planting material of known sex for commercial cultivation or build-up of plant material of a specific sex-type. For this purpose enhanced axillary shoot proliferation from nodal segments are generally used as explants (Rai, *et al.*, 2012). Most of the time MS media is used for root and shoot proliferation with varied concentrations of growth substances like auxins and cytokinins. As compared with other propagation methods, micro-propagation is most important in this crop for raising true to type plants of selected line. But this is quite expensive and need to develop a protocol for better success of regeneration.

Sowing/Planting

Sowing of seeds is done at a depth of 2 cm to 3 cm directly into the soil after treating with chemicals like sulfuric acid or any germination inducing growth regulators *viz.* gibberellic acid (GA_3) at 200 ppm and 2,4-D at 100 ppm (Bhuyar *et al.*, 2000). Combination of both scarification and hormonal treatment of 10-15 minutes gives better result of germination. In case of stem cuttings, pre sprouted cuttings of both female and male plant are directly planted in the main field in pre-fixed proportion. While tuber propagation, these can directly planted in the main field after treating with any fungicide to avoid rotting of tubers. Mostly whole tuber is used for planting instead of making into pieces. If there is availability of seedling produced through micro-propagation they can also raised directly after undergoing proper hardening process.

Season of Planting or Sowing

This crop can be grown successfully in two seasons in a year *viz.* rainy season and spring-summer season. Rainy season crop is sown during the months of June to July while, in case of spring-summer crop sowing is done during the months of February to March. In tropical hills sowing used delayed up to April month due to low temperatures (Bairava and Rana, 2017 and Khare, 2004).

Irrigation Management

Spine gourd is majorly considered as rainfed crop but for commercial production it can also grown as irrigated crop. Underground tuber is used to have lateral roots, through which it absorbs the water and nutrients. Soon after planting irrigation is given and continued daily for first two weeks. After establishing the crop irrigation can be applied based on the need of the crop. Flowering and fruit setting are the critical stages of irrigation. Removal of excess moisture or stagnated water by providing proper drainage facilities is also important to protect the crop from rotting. Irrigation should stop after completion of the crop, when the vine reaches to senescence. The underground tubers are remain in dormancy during winter and left in the field as it is.

Nutrition Management

Spine gourd is not considered as heavy feeder of nutrients but it demands the soil with high organic matter content. To obtain better returns, maximum organic manures *viz.* FYM at 3-5 kg per plant; vermicompost at 1-2 kg per plant is incorporated into the soil at the base of the plant during starting of the season. It can be suggested to apply nitrogen (N), phosphorus (P_2O_5) and potassium (K_2O) at the rate of 60:30:45 kg per hectare respectively [10]. The total phosphorus and half the nitrogen and potassium are applied as basal dose. Remaining half nitrogen is applied in two splits at 35 and 60 days after vine emergence and half potassium is applied during flowering and fruiting *i.e.* 45-50 days after vine emergence. Based on the deficiency symptoms on the plant secondary and micronutrients are applied as foliar sprays.

Weed Management

Weed management in spine gourd is one of the important operations as it is a high row spaced crop, used to invite huge weed population during early stage of the crop. To avoid heavy weed infestation two or three hoeing should be practiced between the plant rows. As crop emerges and spreads the vines, dense foliage itself will suppress the weed growth. As this crop is shallow rooted, it is better to not recommend chemicals for controlling the weeds.

Pest and Disease Management

Insect pests like fruit flies, green bug, painted bug, epilachna beetle, red pumpkin beetle, grasshopper, cotton grey weevil, tobacco caterpillar and leaf eating caterpillar are usually causing the damage in spine gourd (Singh *et al.*, 2009). Downy mildew, anthracnose, powdery mildew, mosaic and angular leaf spot are common diseases along with nematodes (Saj, 2019).

Fruit Flies (*Bactrocera cucurbitae*)

Maggots causes damage by feeding on the pulp of the fruits and can be identified by oozing of resinous fluid from fruits. Due to this, fruits will become distorted and malformed finally leads to premature dropping. Such fruits are unfit for marketing and consumption.

Management

Collective operations are required to manage this pest *viz.* collect infested and fallen fruits and bum in deep pits, change the sowing date as the fly population is low in hot dry conditions and at its peak during rainy season, expose the pupae by ploughing and turning over soil after harvest and use ribbed gourd as trap crop and apply Emamectin Benzoate 5 per cent SG at 0.4-0.5 g or Acetamiprid 20 per cent SP at 0.1 g or Carbosulfan 25 per cent EC at 1-1.5 ml per liter of water on congregating adult flies on the undersurface of leaves. Use attractants like citronella oil, eucalyptus oil, vinegar (acetic acid), and lactic acid to trap flies. Use of baits *viz.* Peel of oranges or cucumbers or their fruit pulp, 100 ml ammonia or cow urine, 0.5 litre of water; all mixed well and allowed to stand overnight. For ready-to-use bait, mixture is diluted with 15 litre of water. In another method 6 ml yeast extract, 0.5 g sodium sulphite (Na_2S), 1 litre of water is used (Stoll, 2000). Insecticides can also be used in a mix with attractants like cue lure and methyl eugenol. This is a technique called Male Annihilation Technique (MAT) and consists of many bait traps (16 traps) throughout the field. The mixture can be applied in traps or other devices. Insecticides can also be mixed with protein bait, which controls both, male and female fruit flies. A hydrolysed protein insecticide mix is applied on spots (spot technique) on the crop canopy [37]. Spraying the soil under trees or crops with Metharhizium anisopliae or other entomopathogenic fungi may be helpful in reducing fruit fly larvae in the soil. It also requires that the fungal product used exhibits high effectiveness against fruit fly.

Beetles and Weevils (Epilachna Beetle, Red Pumpkin Beetle and Cotton Grey Weevil)

These insects feed on foliage by making characteristics holes on the leaves. In severe stage total foliage will be eaten and causes serious loss to crop.

Management

1. Collection and destruction of adult beetles
2. Removal of alternate hosts from field and bunds
3. The predator reduviid bug, *Rhinocoris fuscipes* (Fab.) is a larval parasites can be released for parasitization (Rajagopal and Trivedi, 1989).
4. Spraying with effective insecticides like Chlorpyriphos 20 per cent EC at 2 ml or Profenofos 50 per cent EC at 2 ml or Emamectin Benzoate 5 per cent SG at 0.4 g per liter of water.

Caterpillars (Tobacco Caterpillar and Leaf Eating Caterpillar)

These insects larval stage act as defoliators feeds on foliage of the plant and can cause losses up to 100 per cent. These are polyphagous in nature can remain feed on multiple hosts for surviving its population.

Management

1. Monitoring can be done with the help of pheromone traps of up to 10 per hector.
2. Collection and destruction of caterpillars in initial stage of infestation.
3. Providing bird finches in the field
4. Removal of alternate hosts
5. Spraying with chemicals *viz.* Chlorantraniliprole 18.5 per cent SC at 0.3 ml or Emamectin Benzoate 5 per cent SG at 0.4 g or Fenpropathrin 30 per cent EC per liter of water will effectively manage the caterpillars.

Bugs (Green Bug, Painted Bug)

Damage caused by bugs ranges from many small white spots on the leaves to distortion or destruction of plant tissue. Adults fly readily and are often gone before symptoms appear.

Management

1. Identification and destruction of nymphs and adults.
2. Spraying with reflecting agents like neem oil or neem seed kernel extract.
3. For killing the insects spraying with Thiamethoxam 25 per cent WG at 0.3 g or Imidacloprid

4. 17.8 per cent SL at 0.3 ml or Acetamiprid 20 per cent SP at 0.1 g per liter of water can be used.

Mildews (Powdery mildew and Downy mildew)

The first sign of powdery mildew is pale yellow leaf spots. White powdery spots can form on both upper and lower leaf surfaces, and quickly expand into large blotches. The large blotches can cover entire leaf, petiole and stem surfaces. When powdery mildew infects the majority of the foliage, the plant weakens and the fruit ripens prematurely. While, in downy mildew lesions are formed and that are limited by the leaf veins. During periods of leaf wetness from dew, irrigation or rainfall, incipient lesions can become conspicuously water-soaked. This is the earliest symptom produced by the disease, but will disappear as moisture dissipates. Early lesions are light green in appearance and become chlorotic and finally necrotic as host plant cells die. Severe infection results in leaves that are completely dead and curled up. This symptom has been described as "wildfire" as the leaves appear to be burned.

Management

1. Avoid excess nitrogen application
2. Spraying with fungicides *viz.* Azoxystrobin 23 per cent SC at 0.6 g or Chlorothalonil 75 per cent WP at 1.5 g or Hexaconazole 2 per cent SC at 2.5 ml or Propiconazole 25 per cent EC at 1 ml per liter of water will serve the purpose.

Mosaic

It is a viral disease rarely observed in spine gourd, it can be managed by selecting disease free planting material along with frequent sprays with systemic insecticides like Thiamethoxam 25 per cent WG at 0.3 g or Imidacloprid 17.8 per cent SL at 0.3 ml per liter of water.

Use of Plant Growth Regulators

It is possible to induce male flowers in female plants or vice-versa can be possible with the help of certain growth regulators GA_3 at higher concentrations (3000 ppm) can induce male flowers and same at lower concentration (50 ppm) is used for female flower induction. This forcible flower induction can be used for seed production purpose. 2, 4-D and GA_3 is used for increasing germination percentage of the seeds (Bhuvar, 2000). Application of 500 mg/l $AgNO_3$ on 30 days old female plants produced the maximum proportion of induced hermaphrodite flowers (Sanwal, *et al.*, 2011 and Megharaj *et al.*, 2017).

Harvesting and Yield

Harvesting of the fruits starts from 70-80 days after sowing if seed is used for growing. While in case of vegetative methods it takes 55-60 days after sprouting. In general immature fruits of light green to dark green are preferred for harvesting. Harvesting can be done manually in every 3-4 days interval and it continues for 30 to 40 days. The yield potential of the spine gourd varies from 50 - 100 quintals per hectare based on the type of variety and management practices.

Postharvest Management and Marketing

The immature and well developed fruits are the economic part of the plant. These fruits are having better keeping quality as compared with cucumber and bitter gourd. We can increase the storage life of these fruits by storing at 8-10°C up to 10-15 days. Though spine gourd is fetching better price in the market, peoples in some parts of the country are not much familiar about this vegetable. So, there is a need to popularize this vegetable among the consumers, as it is having huge nutritional and medicinal benefits.

REFERENCES

Aberoumand A, (2011) Screening of less known two food plants for comparison of nutrient contents: Iranian and Indian vegetables, Functional Foods in Health and Disease, 10: 416–423.

Anjana M, Swathi V, Ramya Sai A, Divya N, and Sunisha, Y (2019) A Review on *Momordica dioica* Fruits. J Adv Plant Sci 3(**1**): 1-5.

Anjaria J, Parabia M, Bhatt G, Khamar R (2002) Natural heals: A glossary of selected indigenous medicinal plants of India.Sristi innovations 2nd edition.Ahmadabad: 35.

Bairava SK, and Rana MK, (2017) Vegetable crop science. In MK, Rana (Ed.). Spine gourd. London, United Kingdom, Taylor and Francis Ltd: CRC Press.

Bandyopadhyay S, Mukherjee SK (2009) Wild edible plants of Koch Bihar district, Natural Product Radiance, 8(**1**), pp. 64–72.

Baratakke RC, Patil CG, Poornima B, and Sankannavar SH, (2013) Molecular tool for sex identification (female) in *Momordica dioica* Roxb. with reference to medicinal values. Int J Res Ayur Pharm 4(**4**): 487-490.

Bharga V, (2017, December). Spine gourd: A miraculous vegetable. Kerala Karshakan e-journal. 5(**5**): 24-25.

Bhuyar SA, Dod VN, and Gholap SV, (2000) Studies on the effect of growth regulators on seed germination in spine gourd (*Momordica dioica* Roxb). Orissa J Horti 28(**1**): 56-58.

Chopra RN, Nayar SL, and Chopra ZC, (1956) Glossary of Medicinal Plants of India. Publication Council of Scientific of Industrial Research, New Delhi. 169.

Dubey AK, Srivastava JP, and Singh NP, (2006) Studies on floral biology of spine gourd (*Momordica dioica* Roxb.). In I International Conference on Indigenous Vegetables and Legumes. Prospectus for Fighting Poverty, Hunger and Malnutrition 752 (pp. 453-458).

Duthie JF (1965) Flora of the upper Gangatic plans. Botanical survey of India. Calcutta. 1: 89-91.

German Agency for International Cooperation (GIZ). (n.d.). Fruit fly Management. Pp 6.V. https: //www.asean-agrifood.org/?wpfb_dl=160.

German Agency for International Cooperation (GIZ). (n.d.). Fruit fly Management. pp7. https: //www.asean-agrifood.org/?wpfb_dl=160.

Hooker JD (1961) The Flora of British India, vol. 2, Reeve Co, Kent, UK.

Jadeja BA, Odedra NK, Odedra KR (2006) Herbal remedies used for haemorrhoids by tribals of Saurashtra, Gujarat. Indian Journal of Traditional Knowledge, 5(**3**). 348–352.

Jain A, and Singhai AK (2009) Effect of *Momordica dioica* Roxb, on gentamicin model of acute renal failure. Nat Prod Res 2: 1-10.

Khare CP (2004), Encyclopedia of Indian Medicinal Plants. Springer, New York: 104.

Khulakpam NS, Singh V, Rana DK (2015), Medicinal importance of cucurbitaceous Crops. *Int Res J Bio Sci* 4(**6**): 1-3.

Kirtikar KR, and Basu BD (1999). Indian Medicinal Plants. International Book Distributors, Dehradun. 2: 1129-1135.

Kumar U, Prajapati ND (2003). Agro's Dictionary of Medicinal Plants. Agrobios (India), Agrohouse Jodhpur, India, pp.216.

Megharaj KC, Ajjappalavara PS, Revanappa, Manjunathagowda DC, and Bommesh JC, (2017) Sex manipulation in cucurbitaceous vegetables. Int J Curr Microbiol App Sci 6(9): 1839-1851.

Momordica dioica. (2020). In *Wikipedia*. Retrieved from https: //en.wikipedia.org/wiki/Momordica_dioica.

Nadkarni KM (2004) Medicinal plants of India. Reprint publication Dehradun. India. 236-237.

Oommachan M (1977) The flora of Bhopal. JK, Jain brothers, Bhopal. India. 180-181.

Oudhia P, Ankol (Alangium) based Indigenous Herbal Medicines for Bleeding Piles (Hemorrhoids): Pankaj Oudhias Medicinal Plant Database. http: //www.pankajoudhia.com/.

Ponnusamy J, Arumugam B. Spine gourd: An underutilized Cucurbitaceous vegetable. Kerala Karshakan e-journal. Pp45-47. https: //www.researchgate.net/publication/330497782.

Rai GK, Singh M, Rai NP, Bhardwaj DR, Kumar S (2012) *In vitro* propagation of spine gourd (*Momordica dioica* Roxb.) and assessment of genetic fidelity of micropropagated plants using RAPD analysis. *Physiology and Molecular Biology of Plants*, 18(**3**): 273-280.

Rajagopal D, Trivedi TP (1989) Status, bioecology and management of Epilachna beetle, *Epilachna vigintioctopunctata* (Fab.) (Coleoptera: Coccinellidae) on potato in India: A review, Tropical Pest Management, 35: 4, 410-413, DOI: 10.1080/09670878909371418.

Rajasree RS, Sibi PI, Francis F, William H (2016) Phytochemicals of Cucurbitaceae family-A review. *International Journal of Pharmacognosy and Phytochemical Research*, 8(**1**): 113-123.

Ram D, Banerjee MK, Pandey S, Srivastava U, (2001) Collection and evaluation of Kartoli (*Momordica dioica* Roxb. Ex. Willd.). *Indian Journal of Plant Genetic Resources*, 14(**2**): 114-116.

Ram D, Kumar S, Verma A, and Rai M, (2004) Variability analysis of underutilized nutritive vegetable Kartoli: Indian collection. Report-cucurbit genetics cooperative, 27: 66.

Saj S, (November 21, 2019). Spine gourd farming (Kakrol/Kantola); Information and guide. https: //www.farmnewsindia.in/2019/11/21/spine-gourd-farming-kakrol-kantola-information-and-guide/.

Sandilya VK, Ram N, Pradhan V, Kumar R, and Tiwari JK, (2019) Pollination and fruit setting mechanism in spine gourd (*Momordica dioica* Roxb). J Pharm Phytochem 8(**6**): 685-690.

Sanwal SK, Kozak M, Kumar S, Singh B, and Deka BC, (2011) Yield improvement through female homosexual hybrids and sex genetics of sweet gourd (*Momordica cochinchinensis* Spreng.). Acta physiologiae plantarum, 33(**5**): 1991-1996.

Sastri BN, (1962) Wealth of India-Raw Materials. Council of Scientific and Industrial Research, Delhi, 406-407.

Satyavati GV, Raina MK, Sharma M, (1987) In Medicinal plants of India, ICMR New Delhi. (**1**): 327.

Singh PK, Singh AK, Singh HM, Kumar P, and Yadav CB, (2009) Insect-pests of spine gourd (*Momordica dioica* Roxb.) and efficacy of some insecticides against the epilachnid beetle, *Henosepilachna vigintioctopunctatu* (F.), a serious pest of this crop. Pest Management and Economic Zoology, 17(**1**): 85-91.

Stoll G (2000) Natural Crop Protection in the Tropics. Margraf Verlag 2000, Weikersheim, Germany.

Talukdar SN, and Hossain MN (2014) Phytochemical, phytotherapeutical and pharmacological study of *Momordica dioica* evidence-based complementary and alternative medicine. 1-11. http: //dx.doi.org/10.1155/2014/806082.

Index

A

Abelmoschus spp., 35
Abiotic stress 14
African group 9
Alternaria cucumerina 154
Alternaria leaf blight 154
Amaranthus 112
Angular leaf spot 154
Anthracnose 69, 155
Antioxidants 53
Arka Neelachal Shree 163
Artificial induced parthenocarpy 125
Artificial inoculation 22
Artificial screening 23
Asiatic group 10
Aulacophora foviecollis 153

B

Backcross 41, 84
Bacterial canker 76
Bacterial diseases 74
Bacterial speck 76
Bacterial spot 77
Bactrocera cucurbitae 152, 167
Beetles 168
Bitter gourd 31
Botany 3, 148, 162
Bottle gourd 30
Breeding 1, 19, 35, 47, 61
Breeding approaches 53, 128
Breeding cycle 116
Breeding methods 41, 84
Brinjal 136
Bud necrosis 80

C

Capsaicinoids 53
Capsicum 137
Caterpillars 168
Classification 5
Climatic requirements 150, 164
Colletotrichum lagenarium 155
Controlled environmental Chamber 108
Conventional breeding 105

Cotton grey weevil 168
Crop improvement 108
Crop production 149, 163
Cucumber 1, 22, 133
Cucumis 1
Cucumis melo var. *momordica* 145
Cucurbits 19
Cultivars development 105
Cultivation 145
Curcurbitaceae 1
Cytotaxonomy 9

D

Detached leaf assay 23
Dietary fiber 52
Disease resistance 61
Downy mildew 19

E

Early blight 66
English cucumbers 9
Epilachna beetle 168
Etiology 21
Etymology 3

F

Fast trait stacking 114
Field cucumbers 9
Field preparation 151, 164
Floral biology 4, 162
Flower 4
Food security 105
Fruit flies 167
Fruit fly 152
Fungal pathogen 63
Fusarium wilt 69

G

Gene pyramiding 61
Genera 1
Genetic engineering 48, 55, 130, 133
Genetic linkage map 29
Genetic parthenocarpy 125
Genetic resources 1
Genome editing 105
Genotyping 105
Germplasm 53
Gherkins 9
Glasshouse speed breeding conditions 108
Gray leaf spot 72
Green bug 168
Greenhouse design 107

H

Habitat 3
Harvesting 155, 169
Heterosis breeding 85
High throughput phenotyping 105
Hybridization 54
Hypersensitivity 22

I

Improved varieties 149, 163
Improvement 1
Indian cauliflower 56
Indiginous cucurbitacous vegetable crop 159
Indira Kankoda I 163
Interspecific hybridization 41, 54, 85
Introgression 56
In vitro propagation 165
Irrigation 152, 166

K

Karyotypic characters 10
Konkan Madhur 150

L

Late blight 63
Leaf curl disease 79

Leaf disc method 22
Leaf eating caterpillar 168
Leaf minor 153
Leaf mold 68
Leafy vegetable crops 112
Legume crops 109
Lemon cucumber 9
Liriomyza spp 153
Low-cost homemade growth rooms 108
Lycopene 53

M

Malnutrition 49
Mapping parthenocarpy 134
Marker assisted selection 61
Marketing 156, 170
Micro-propagation 165
Mildews 169
Minerals 51
Minor cucurbit 145
Mitotic chromosomes 10
Moisture 21
Molecular approaches 86
Molecular breeding 129
Molecular markers 74, 75
Momordica dioica 159
Mosaic 169
Multipartite plant viruses 37
Muskmelon 28
Mutation 41, 55
Mutation breeding 129

N

Nutraceutical 48
Novel parthenocarpy QTLs 133
Nutritional quality 47, 49
Nutrition management 152, 166

O

Okra 35
Oriental-type cucumber 9
Orphan crops 113

P

Painted bug 168
Parthenocarpic gynoecious lines 135
Parthenocarpic vegetables 123
Pea 109
Pedigree method 54, 85
Pest and disease management 152, 167
Phylogenetic affinities 12
Plant growth regulators 130, 155, 169
Planting 166
Plant morphology 3
Plant viruses 36
Pollination 4
Postharvest management 156, 170
Powdery mildew 69, 169
Production technology 145
Propagation 151, 165
Protein 49, 55
Pseudomonas lachrymans 154
Pumpkin 31
Pure line selection 54
Pusa shandar 149

Q

Quality traits 48

R

Rainfall 21
Rapid cycling 114
Red pumpkin beetle 153, 168
Resistance sources 61, 75
Ridge gourd 30
Root and tuber vegetable crops 111
Root-knot nematodes 72

S

Scientific classification 145
Screening technique 21, 42
Seed 5
Seed rate 151
Seed (sexual) propagation 165
Selection 53
Sensitive assay 42
Septoria leaf spot 72
Sikkim cucumber 9
Single seed descent method 85
Smoot-skinned beit alpha types 9
Snap melon 31, 145
Soil 151, 164
Solanum lycopersicum 61
Somaclonal variations 54
South-east Himalaya 9
Sowing 166
Sowing season 151
Spacing 151
Species forming 10
Speed breeding 105
Spine gourd 159
Standard protocol 107
Synthetic amphidiploid species 12
Systematics 1

T

Taxonomic position 2
Taxonomy 1, 14
Temperature 21
Tobacco caterpillar 168
Tomato 61, 112, 131
Tomato mosaic virus 82
Tomato viral diseases 78
Tomato yellow leaf curl disease 61, 78
Transgenic approach 130
Transgenics 85
Tuber cuttings 165

U

Uses 148, 161

V

Varieties 19
Vegetable breeding 48
Vegetable crops 105, 113
Vegetable improvement 117
Vegetables 47
Vegetative (asexual) propagation 165
Vegetative plant 3
Verticillium wilt 70
Vine cuttings 165
Vitamins 50

W

Watermelon 138
Weed management 152, 167
Weevils 168
Wild species 6

Y

Yellow vein mosaic virus 39
Yield 169
YVMV Resistance 35

Color Plates

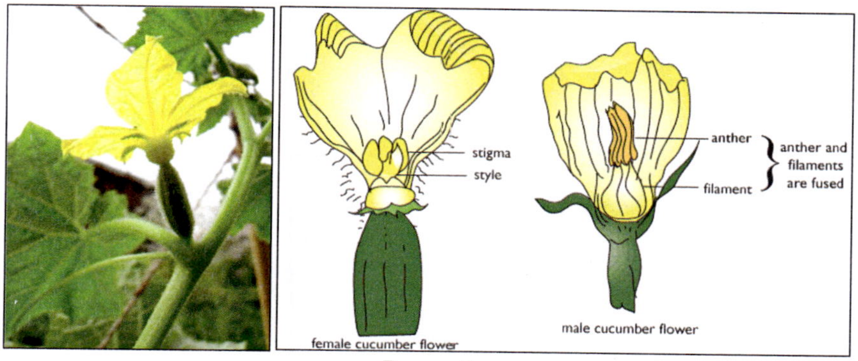

Page: 4

Page: 20

Page: 24

Page: 25

Page: 41

Page: 146

Page: 159

Page: 149

Page: 150

Page: 154

Page: 162

Page: 163-164